MUSIC AND COMMUNICATION

Music and Communication

TERENCE McLAUGHLIN

FABER AND FABER London

First published in 1970
by Faber and Faber Limited
24 Russell Square London WC1
Printed in Great Britain by
Latimer Trend & Co Ltd Plymouth

ISBN 0 571 09352 3

To Eve,
who tolerated many silent evenings
in the interests of music

Contents

9

1 Introduction

Writing about the essential nature of music is very like investigating the essential nature of life—before the investigator has probed very far, the subject has ceased to have any life to examine, and only a corpse is left for study. Music exists in movement and change, and before any part of it can be pinned down for analysis, the music has moved on and is presenting another face. Marvell called music 'the mosaic of the air', but this image seems too static and lifeless to describe the constantly shifting patterns, the sense of growth and movement, the feeling that music is not just a matter of massing one part against another, but an organic process of development.

It is organic, but it is also different from the organic growths of nature such as trees and flowers, or organic movements like the flow of water over the stones of a river bed, or the movement of the sea, in that it expresses human experience. This present book is devoted to an analysis of the possible ways in which music can be expressive and the mechanisms by which a great many people derive such pleasure from the experience so transmitted.

The action of music on human beings, and the means that it uses, form a fascinating and baffling problem. Browning put it well in *Abt Vogler*:

But here is the finger of God, a flash of the will that can,
Existent behind all laws, that made them and, lo, they are!
And I know not if, save in this, such gift be allowed to man,
That out of three sounds he frame, not a fourth sound, but a star . . .

There seems no reason, on the face of it, why music should have any human value at all—the point which Browning makes—or why, indeed, it should ever have been invented. Language, drawing and similar skills have a definite survival value as a means of communication for mundane affairs; building is similarly a useful art. While it is difficult to trace the details of the path by which these skills were developed and refined into the arts of poetry, painting, sculpture and architecture, there is nothing strictly mysterious in the fact that such a thing should have happened.

Music, on the other hand, is not useful, and not even imitative. Musical sounds are not very common in nature, except for bird song and tricks of wind and water, and the patterns of relatively pure and sustained vibrations which are characteristic of music cannot therefore appeal to human beings because of their association with some real object. In fact, unlike the other arts, the raw material of music cannot readily be associated with anything at all except similar musical sounds—and the process must start somewhere. If we follow the lead of some authors and say that the meanings of music are learned from previous musical experience, we are bound to end the trail with some primitive man who derived *some* pleasure from a crude musical activity which could not be associated with any previous similar experience. Yet the pleasure must have been there, otherwise our primitive forebear would have forgotten about the accidental discovery, and the subject of this book would never have existed.

The problem of the complete uselessness of music for 'ordinary' purposes or for survival has given rise to many theories of its genesis which are more remarkable for their ingenuity than for any illumination that they cast on the facts.

One theory proposes that primitive man imitated the mating cries of animals, or the songs of birds. Unfortunately, the theory does not explain the crucial point, why he should take any pleasure in making these sounds, or why he should want to repeat the experiment. Apart from this, the imitative theory should lead to the conclusion that the most primitive types of music still in existence should have a close resemblance to the sounds of birds and animals, with a move away from this as music-making becomes more sophisticated. This is not so in fact. Primitive music as it exists today in isolated communities seems to place much value on rhythmic devices which have no place in the calls of animals, while the imitation of birds only arises in sophisticated stages of composition, as in Bishop's 'Lo Hear the Gentle Lark' or the *Reveil des Oiseaux* of Messiaen: the noble savage no doubt takes birds for granted, as do most country people today, and cannot be bothered to imitate their chatter.

Another theory, which also begs the question, is that music arises from the lalling of children. Here again, while most children do lall at some time or other, we have no help from the theory in deciding why they *like* to do it. Révész[1] points out how early in life children adopt the musical elements of the tunes which they hear from their parents and other elders: even at two years old the lalling of infants has some of the characteristics of the popular music of their native country, just as the almost incomprehensible prattling of these young children still has the customary intonations of their parents' language. Small Dutch children, for example, still reproduce the guttural Dutch 'g' however shrill their voices. In general it seems that there is no universal language of children's music, and no spontaneous generation of music by them: they merely copy their elders.

Utilitarian theories of the genesis of music are similarly deficient in supplying any real reason why human beings should actually want to make music. Herbert Spencer[2] tried to trace a development of music from speech gestures, the intonations of excited speech being gradually disengaged from the words and transferred to song: 'Those vocal peculiarities which indicate excited feeling, are those which especially distinguish song from

---

[1] *Introduction to the Psychology of Music*, trans. de Courcy, Longmans Green, London, 1953.
[2] 'The Origin and Function of Music', *Fraser's Magazine*, 1857.

speech.' This theory drew upon itself a host of critics, including Darwin, Edmund Gurney and Ernest Newman: while it certainly accounts for the 'language' of volume and tone colour in musical rhetoric, it does not well account for the origins of melody or harmony. Darwin supposed the contrary theory in *The Descent of Man*, that speech did not precede music, but was rather derived from it. Singing, for Darwin, was one of the means of attraction in mating, and speech came later as a development of this activity. The controversy between Spencer and Darwin is covered in more detail by Peter Kivy;[1] for the present purpose it is sufficient to note that neither Darwin nor Spencer made any attempt to answer the question *why* music was attractive in the lives of primitive men or their descendants. A similar objection applies to Révész's own theory of 'calling signals' as the beginnings of musical developments. He points out correctly that for communication over long distances, speech is inadequate, and various musical calls—coo-ee, yodelling and so on—have to be used instead. He then goes on to point to the pleasure which can be derived from making these calls even in the absence of a need for communication and traces the development of singing from this source. But whence the pleasure derives is not stated: the theories tend to move in ever-decreasing circles.

Wherever we turn, the origins of music remain mysterious because of the dichotomy between the human experience which constitutes the content of the art, and the inhuman, mathematical *means* of expression. The lack of any overt external association marks off music from other arts. When Lear is raving on the stage, we probably see him as an archetypal figure of the human condition when it has reached the wilderness of despair, but even if we do not, he is still a recognizable old man who has been ill-treated and left out in some extremely unpleasant weather. When Rembrandt paints a portrait or Turner a landscape, we can see that here is an interesting face or a wild

---

[1] 'Herbert Spencer and a Musical Dispute', *Music Review*, Vol. 23, 1962, p. 317.

stormy sky, even if we discern nothing else in the picture at all. Even architecture cannot be divorced from the practical usefulness of the building, and this can sometimes be in itself a moving experience. I have stood under the Pont du Gard, that great Gallo-Roman aqueduct which once carried water to Nîmes, and felt a great sense of wonder and pleasure in the purpose of the structure, as well as in its proportions and grace. Most of the great works of architecture which one sees are monuments to superstition, like the Pyramids, or temples, or cathedrals, or they are monuments to the useless wars which have troubled mankind for so long. Here was a fine and majestic structure dedicated to the civilized purpose of bringing fresh water to a great city.

Some element of extraneous association of this kind must enter into every art which deals with material taken from nature. Even the most abstract of pictures, for example, may contain patterns which are significant to us because they correspond to some real object which is emotionally important— sexual or otherwise, according to one's favourite theories—and words cannot help having strong everyday associations, however idiosyncratic their arrangement in poetry or prose. Music has no such roots in the real world: this is probably why Schopenhauer referred to it as the true voice of the Will (the Life-Force, the great motivating quality in Schopenhauer's philosophy) while all the other arts were only representations of aspects of the Will. Music expresses no simple associative information and the amount of imitative material, the sound of the cuckoo, of sheep, of motor horns, and so on, which can be used is negligible. Yet most people would still contend that it expresses something which is deep and valuable and which can be communicated from the composer to the listener in such a way that it conveys some aspect of the composer's subjective thought.

Some writers, of course, deny that this is the case. To them, music expresses nothing except itself and our pleasure in it consists only in recognizing the beauty of the proportions or of the technique. These ideas go back as far as Pythagoras, who saw music as the voice of universal harmony, in which we are allowed to share, at times, at a due distance, by the grace of Aoide and the composer.

This theory sees the composer as interpreter and priest of the music of the spheres:

> *. . . then listen I*
> *To the celestial Siren's harmony,*
> *That sit upon the nine infolded spheres,*
> *And sing to those that hold the vital shears,*
> *And turn the adamantine spindles round,*
> *On which the fate of Gods and men is wound.*
> *Such sweet compulsion doth in music lie,*
> *To lull the daughters of Necessity,*
> *And keep unsteady nature to her law,*
> *And the low world in measured motion draw*
> *After the heav'nly tune, which none can hear*
> *Of human mould, with gross unpurg'd ear . . .*
>
> MILTON, 'Arcades', I

Plato tells us that a muse, or siren, sat on each planet, to sing the song peculiar to that body's own motion, yet harmonizing with that of the others. Maximus Tyrius notes that the motions of the planets must create sounds, and as the motions are in simple mathematical proportions, the sounds must harmonize.

The idea of music as 'audible arithmetic', registering for humankind (however imperfectly) an ideal of proportion which is universal and superhuman, is very comforting to a certain type of mind, and this view has had champions in every age. The pantheist detects in music one of the voices of the universal harmony, as Wordsworth in 'On the Power of Sound':

> *By one pervading spirit*
> *Of tones and numbers all things are controlled*

and others may say that music is the voice of God and therefore needs no further exegesis.

For most of us, however, music is the voice of man, and a

remarkably subtle and expressive voice. The 'emotions' expressed by music are not easy to translate into any other form, and harder still to define: and they have depths which do not seem to be expressed properly by any other art and by words least of all. This helps us to define the questions which any adequate theory of musical communication must answer. Not only must we find a mechanism by which music can express something to human beings, but we must also make sure that it is a sufficiently flexible and comprehensive mechanism to explain the depths and subtleties of musical expression. I have written this book in the firm belief that music expresses human experience, and that it is possible to present a coherent, if not complete, theory of the mechanism by which the experiences are communicated, and what kind of experiences they are. To do this it is necessary to analyse the art of music into its component parts so that these can be compared with their possible effects on our brains: the next chapter is concerned with the elements of music analysed in this way.

2 The Elements of Music

To analyse the effects of such a complex art as music, with such varied qualities and so many different facets, it is necessary to analyse the music itself, to determine the main elements which play a part in its constitution, and to see what effects these have, singly and in combination.

Music differs from mere noise in that it consists of vibrations in the air, or combinations of vibrations, which remain constant long enough for the ear to be able to distinguish them as entities or units, the *notes*. Noises may well contain exactly the same vibrations, but these are sustained for so short a time that the ear does not have the opportunity to characterize them as notes. For example, the ringing sound produced by a wet finger on a glass is musical because it is sustained. The same sound may well occur in the crash of the same glass breaking, but the ear will not have a chance to characterize it as a note before it has been succeeded by other sounds. There are boundary cases, of course. The note of a timpanon contains an identifiable note at a definite pitch, mixed up with a great deal of unclassifiable noise. There is the thump of the stick on the skin of the drum, which sets up very transient vibrations, mixed with the steady vibration of the skin itself, which continues long enough for us to recognize the pitch. The note of a side-drum contains more noise and less musical sound, and it is difficult to say that it has any definite pitch, but we can at least recognize that it is higher than that of a tenor or bass drum.

To state the obvious, another quality of music, as opposed to simple musical sound, is that there must be some variation in the notes used—a monotone is not music. As the notes alter, there are consequent relationships set up between them, relationships of pitch, dynamics, duration, and so on, and we may follow Hindemith's phraseology in dividing these relationships into *tensions* and *resolutions*. The tensions are the relationships which are less pleasant to the ear, the resolutions those more pleasant. The art of music lies in organizing these tensions and resolutions into patterns which are significant to the listener.

The tension/resolution patterns which are normally employed in music occur in the three modes of *pitch, time* and *volume*, with further effects possible in *timbre* and *texture*. An excellent analysis of the use of such elements, especially that of pitch, is given in Deryck Cooke's *The Language of Music*,[1] to which the following sections of this book owe much. The notes in this present chapter are only intended to outline the main characteristics of such patterns: a rich collection of examples is contained in Mr Cooke's musical lexicon, to which the interested reader is recommended.

Pitch Tensions

Tensions between notes of different pitches can occur in two ways: there is the element of sheer distance of rise or fall, conveniently described as *intervallic tension*, and there is the interaction of the notes as members of a harmonic series which are conveniently called *tonal tensions*. In both cases the notes may be sounded separately or together, so as to give *melodic* or *harmonic* combinations.

Intervallic tensions have an obvious affinity with the physical effort of singing the notes. Peter Grimes expresses the physical

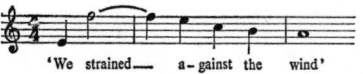

'We strained—— a-gainst the wind'

reality of the strain by the parallel between the muscular effort on board Grimes's boat and the strain on the muscles of the

[1] *The Language of Music*, Oxford, London, 1959.

singer's vocal cords, even without the added harmonic tension of the minor ninth. A minor second would have had the same harmonic sense, but it would hardly express the phrase so vividly. Similarly the rising major sevenths in Elgar, Walton and Bliss have given a sense of almost ferocious aspiration to their music which has left its mark on the whole of twentieth-century British composition. A falling phrase of one semitone would scarcely have left this impression.

On the 'downward' path we have only to think of the violent leaps which mark the final doom of Don Giovanni in the second act of Mozart's opera.

'Ah! tem - po più non v'è!'

Intervallic tensions do not, of course, have to be as dramatic as this to achieve an effect of tension. There is, in all the modes of tension which we shall consider, a question of relativity. In the tense, turbulent atmosphere of Walton's First Symphony, for example, only rising sevenths, fortissimo, will register with the listener as an additional point of stress. In the quiet melisma of a folk-song, normally moving along successive notes of the scale, a rise or fall of as little as a third or fourth can create a feeling of tension by contrast with the rest of the melodic line. In 'She Moved Through the Fair', for example:

the falling fourths are quite enough to create a climactic point in the quietly-moving phrases.

In most cases, higher notes, and leaps upward, mean added tension. With the human voice, the muscles of the vocal cords tighten; with a brass instrument, the player's lips are more stretched; with woodwind, the player blows harder or more acutely. On the piano keyboard and stringed instruments, there is no sense in which high pitch creates physical tension and the leap itself becomes more of a point of stress than the direction, but, possibly by analogy with the voice, most people appear to feel the upward movement as passion, aspiration, protest, triumph, and other positive, 'out-going', tensions, whilst a fall in pitch is associated with the expression of acceptance, resignation, comfort rather than out-and-out pleasure—the 'in-coming' tensions.

Tonal tensions arise from the fact that the combination of some notes in the scale seems pleasant to the ear, while other combinations are unpleasant. The theory of music as practised in the western world is almost entirely founded on the tension/resolution patterns of such concords and discords and their relationships as expressed in harmony, counterpoint, key-structure, and so on. There is no body of theory to advise composers on the best procedures for rhythmic or dynamic organization, but a vast collection of writing on harmonic matters. It is the more surprising, therefore, that so little is known about the real factors which contribute towards making an interval pleasant or unpleasant to the ear. There is not even agreement about the degree of concordance or discordance of some of the combinations.

The Pythagoreans had a simple and elegant theory which is unfortunately untenable. Pythagoras observed that the ratios of the lengths of strings, or other vibrating materials, followed simple mathematical relationships with the sounds they produced. Thus, if a string was halved in length, it produced the octave of its original note; if it were reduced to two-thirds, it produced the fifth above its original note; reduced to three-quarters, it produced the fourth above, and so on. The tradition tells us that Pythagoras was passing a blacksmith's shop and realized that the musical clinking of the hammer varied in pitch according to the lengths of the pieces of iron being worked.

Whatever the origin, the Pythagorean observations were correct. If a violinist moves his finger up the finger-board of his instrument so as to halve the length of the string between the bridge and the nut, he will obtain a note an octave above the pitch of the open string. If two organ pipes are constructed, one 50 per cent longer than the other, they will sound the interval of a fifth when played together. We now know that the relationships are due to the rate of vibration of the string or column of air—a' in the treble clef vibrates at 440 cycles per second, e'' a fifth above vibrates at $440 \times \frac{3}{2} = 660$ c.p.s., a'' an octave above vibrates at $440 \times 2 = 880$ c.p.s., and so on. Other intervals in the scale were found by the Pythagoreans and others to have the following ratios:

Unison	1:1	(These ratios are in fact those calcula-
Major Second	8:9	ted by Zarlino, about 1560. The Py-
Major Third	4:5	thagorean system had slightly different
Perfect Fourth	3:4	values for the major third, sixth and
Perfect Fifth	2:3	seventh. A full discussion of this is
Major Sixth	3:5	given by Alexander Wood.[1])
Major Seventh	8:15	
Octave	1:2	

It will be seen that the ratios of the concordant intervals, the perfect fourth and fifth and the octave, can be expressed in small figures, while the discordant second and seventh have more complicated ratios. In the true Pythagorean system the ratio of the major seventh was in fact 128:243, which is even more complex than 8:15. To the early mathematicians who discovered these ratios, the underlying arithmetical simplicity was sufficient to 'explain' the pleasantness of concords and the harshness of discords. The mind delighted in such mathematical elegance, even if not consciously aware of it. Leibnitz, the great seventeenth-century mathematician, wrote '. . . music is a hidden arithmetical exercise for the soul in which the soul counts without being aware of it. . . . Although the soul is not aware of this counting, it nevertheless feels the effect of this unconscious counting, that is, consonance produces a pleasant sensation, dissonance an unpleasant one, as the natural consequence.'

There are two great objections to this theory of arithmetical

[1] *The Physics of Music*, 6th edn., Methuen, London, 1962.

pleasure. One is that, under other circumstances, the soul does not seem at all concerned about the arithmetical simplicity of the objects of its pleasure. In the world of the visual arts, for example, there is a tradition, reasonably well-founded, that the ideal proportions for two lengths (the sides of a frame, the distribution of the key points in a picture, etc.) are in the ratio known as the *Golden Mean* or *Golden Section*. This ratio is by no means simple, it is, in fact, what the mathematician calls incommensurable, as it cannot be expressed exactly by any fraction or decimal however many figures are taken. (It is, in detail, the ratio of $1:\dfrac{1+\sqrt{5}}{2}$ or $1:1.61803398\ldots$ to an infinite number of decimal places.) It seems odd that such very different ideas about the ideal proportions should exist in the mind if they are really fundamental to our nature.

The second objection is that intervals can be expressed by very different ratio figures without any practicable effect on their actual sound. Zarlino's major third, quoted above, was expressed as the ratio of 4:5, which we take to be fairly concordant. Pythagoras expressed it as 64:81, which would suggest a violent discord. Yet the actual difference between the two expressions is the *comma of Didymus*, about 5·4 savarts or just over a tenth of a tone. It is very doubtful whether this difference is really heard as the alteration of a fairly concordant chord into a very discordant one, and many people would notice no significant difference at all between Pythagoras's and Zarlino's major third.

Probably the largest step forward in the study of concord/discord relationships was taken by the great nineteenth-century physicist and physiologist H. von Helmholtz. In his *Sensations of Tone*[1] he summarized the whole theory of musical sounds as known up to his day, and added to this probably a greater body of knowledge from his own work. There is hardly an aspect of musical theory, acoustics, the physics of sound waves, or the behaviour and structure of the ear which has not been enriched

[1] *Sensations of Tone*, trans. Ellis, Longmans, London, 1875.

by Helmholtz's theoretical perception or experimental skill. On the subject of intervals, he pointed out, firstly, the relationship between dissonance and *beats*. When two sources of sound are vibrating, they are in fact sending out waves of alternate compressions and rarefactions in the air, the distance between each compression being the *wave-length* of the sound. If the rates of vibration are exactly the same, two compressions and two rarefactions will always occur together, so that the ear receives a sound which is the sum of the intensities of the two sources.

If, now, one of the vibrating sources begins to vibrate slightly slower than the other, the compressions which were 'in phase' and simultaneous, will begin to get out of step. Soon a point will be reached where source One is producing a compression when source Two is producing a rarefaction, and the total effect will be that the ear receives far *less* sound than it did from either of the two sources separately. Later the vibrations will get a whole cycle out of step, and compressions and rarefactions will again occur together, to be followed by another diminution in sound, and so on. Thus the total effect of our two vibrations with a slight difference in frequency is to produce a steady note, with, superimposed upon it, a much slower waxing and waning of the sound. This is known as a *beat*, and it can easily be shown that the frequency of beats, per second, is equal to the difference between the frequencies of the two sources, also measured as cycles per second. Thus, if we have two perfectly-tuned *a'* tuning-forks, they will together give a clear vibration of 440 c.p.s. without beats. If one is slowed down, as with a weight or a piece of wax on its prongs, to 439 c.p.s., there will be a beat of one per second, as a sort of slow warble. At 430 c.p.s., the slower fork will make ten beats per second with the faster, and so on.

If we imagine two notes gradually drifting out of tune with one another in this way, the number of beats per second will, of course, steadily increase. Subjectively we would notice, at first, a degree of difference in the notes which we could tolerate as mis-tuning, then a sense of genuine dissonance, then, finally, the dissonance would get less as the interval approaches a minor third and then a major third, and so on to the almost completely concordant fourth and fifth. Helmholtz first considered that the origin of dissonance as a subjective phenomenon was that the ear was somehow disturbed by beats of a certain frequency or range of frequencies. For instance, if beats at around thirty per second were disturbing for the ear and therefor unpleasant, we should be able to explain the discomfort caused when b' (493·9 c.p.s.) and c'' (523·2 c.p.s.) are sounded together. Unfortunately this cannot be the whole story, because, as pitches get lower, a beat of 30 c.p.s. corresponds to a wider interval, thus the chord C-G below the bass clef has a beat frequency of 32·6 c.p.s. (98–65·4 c.p.s.), yet this is manifestly a concord. Yet it is not such a concord as, say, c'-g' in the treble clef: there is a feeling of thickness and unease about these bass intervals which has given rise to the practical advice, in piano and choral writing in particular, to avoid close-set chords in the bass. Experiments carried out by Stumpf and others, and summarized by Jeans in *Science and Music*[1] show that the intervals assessed as most unpleasant by a panel of observers tended to be wider in the bass than in the treble: the results may be expressed briefly as follows:

Lower Tone		Maximum Dissonance		Dissonance Disappears	
Frequency	Approx. Pitch	Beat Frequency	Interval	Beat Frequency	Interval
96	G	16	$1\frac{1}{3}$ tones	41	3 tones
256	c'	23	$\frac{3}{4}$ tone	58	$1\frac{3}{4}$ tones
575	d''	43	$\frac{5}{8}$ tone	107	$1\frac{1}{2}$ tones
1,707	a'''	84	$\frac{2}{5}$ tone	210	1 tone
2,800	f''''	106	$\frac{1}{3}$ tone	265	$\frac{3}{4}$ tone
4,000	b''''	—	—	400	$\frac{3}{4}$ tone

[1] *Science and Music*, Cambridge University Press, Cambridge, 1937.

This is an indication that dissonance is not just a matter of beat frequency, but neither is it the case that the same intervals are concords or discords whatever their pitch. The truth seems to lie somewhere in between, and as yet there is no satisfactory physical explanation of the dissonant quality of the semitone or other small intervals.

However, assuming the practical dissonant effect of a semitone, which seems to apply all over the normal compass of music except perhaps at very high pitches, it is fairly easy to explain the varying degrees of unrest or repose brought about by the other intervals in the scale. Helmholtz pointed out that all musical instruments, excluding laboratory equipment like the tuning-fork and the electronic oscillator, produce complex vibrations which are not single notes. A string or air column does not have only one vibration over its full length, but can vibrate in halves, thirds, quarters, and so on. Following the rules discovered by Pythagoras, these will vibrate at twice, three times and four times the frequency of the basic note, the *fundamental*. These higher notes are called *harmonics* or *partials*. Thus, if we take A below the bass clef, with a frequency of 55 c.p.s., a practical instrument sounding this note will actually tend to produce a compound of the multiples of 55 c.p.s., thus:

(The asterisked notes are out-of-tune compared with the pitches written. In particular, at partials 21 and onwards, the intervals between the notes become less than semitones.)

The strength of these harmonics, and the ones which occur with the most prominence, vary with the instrument and the conditions, but the first few harmonics in the series can easily be heard on most instruments. This being so, we have to consider that a chord which is nominally written as two notes comes to our ears as two fundamentals plus the attendant set of harmonics for each note. Thus the chord C-G in the bass really comes to us as:

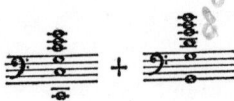

and possibly more harmonics, making a total effect which may be written:

This has no interval in it less than a tone, among the lower harmonics, and this explains the consonant effect of the perfect fifth chord. If we look at the minor third, by contrast, the combinations of harmonics are more uneasy:

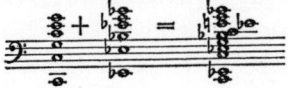

The clash between $e'♮$ and $e'♭$ is very marked, hence the unfinished effect of the minor third and the earlier convention that it was a discord. In general, the Pythagorean idea of consonance can be explained on this basis, because notes which have frequencies in a simple ratio to one another are likely to have a good many harmonics in common, and therefore fewer opportunities for semitone clashes. On the other hand, those with complicated ratios probably have few harmonics in common and are therefore likely to have clashing notes quite low down in the harmonic series—the major seventh, for instance, starts off at a disadvantage, because in, say $c'–b'♮$, the b' immediately clashes with c'', the first harmonic of c'.

Practical tests by experimental psychologists have shown that the order of preference for chords in most people follows the lines predicted from the Pythagorean theory and Helmholtz's explanation of it. There is, however, one main point of difference. Theory would place the 'perfect' intervals, the fourth and fifth, as the most consonant after the octave. In practice listeners seem to find these intervals a little 'empty', and place the richer thirds and sixths higher. C. W. Valentine[1] found the following order of preference: major third, minor third, octave, major sixth, minor sixth, fourth, tritone, fifth, major second, minor seventh, major seventh, minor second. Ortman[2] found a similar order of preference. In a sense there is possible a difference here between consonance and pleasingness: it seems unthinkable that anyone would actually classify the fifth as a discord, but it certainly sounds hollow when used alone.

Given that, despite any other differences of opinion, the sevenths, seconds, and tritone are unpleasant to the ear, we have a basis for analysing the tension/resolution patterns used by composers, whether melodically or harmonically. We can see that the opening theme of the slow movement of Mozart's Piano Concerto in A major, K. 488, owes its grief-laden quality to its emphasis on the minor third and minor sixth degrees of the scale, and even more to the painful seventh and tritone leaps in the melodic line, with a climactic culmination in an intervallic leap from the minor sixth D to the tonic a minor thirteenth below.

Similarly, but in another scale of tension, Holst's climax to Mars in *The Planets* is an orchestral cry of agony *ffff* on full wind and organ. In the already tense and dissonant score of which it forms a part, such an ear-shattering noise is necessary to register the ultimate tension for the climax.

In this connection, there is a school of critical writing, mainly

[1] *The Experimental Psychology of Beauty*, Methuen, London, 1962.
[2] *The Effects of Music*, ed. Schoen, Kegan Paul, Trench and Trubner, New York, 1927.

among those trying to explain 'modern' harmony, which deals with the history of music as if it were a smooth and natural progression of the human ear, from tolerating only unisons and octaves, through *organum*, to a stage of accepting in turn the 'improper' thirds and sixths, then prepared sevenths, unprepared discords, and so on, until it has been trained to accept the final emancipation through Wagner, Strauss and Schoenberg, where the concept of concord or discord ceases to have any meaning. The argument seems to be that we do not *hear* dominant sevenths or unprepared ninths, or for that matter minor thirds or whole-tone chords, as discords any longer, and therefor the composer must find sharper sounds still to awake us to a feeling of tension. It is as if the ear, once ravished by *Electra* or the *Rite of Spring*, could never again recover its innocence. This view is convenient, particularly to the dodecatonal writers, as it can make them appear part of the inescapable development of music (this comes out strongly in Schoenberg's own Textbook of Harmony.)[1] It is also largely nonsense.

First, the harmonic devices used by composers do not develop in a steady flow from more consonant to more dissonant. Bach's style, learned from the great contrapuntists such as Fux, employed harmonic sequences of such complexity that the following generations deliberately simplified their style and diminished their harmonic vocabulary. The logic of Bach's counterpoint, already rich in chromatic passing notes, leads to some quite ferocious discords where he needs a point of particular tension, as in the D minor fugue of the first book of the '*48*', where in two places three successive notes of the chromatic scale are sounded together. This constitutes a discord without parallel in later music until the 'moderns' of the twentieth century. Mozart, particularly in works such as the Requiem which show the influence of Bach, tends to use chromatic tensions which take us into more remote harmonic regions than, say, Beethoven, who relied more on rhythmic tension devices to create his tension-resolution patterns. Purcell created for his

[1] *Harmonielehre*, Universal Edition, Vienna, 1921.

'supernatural' passages harmonic sequences much stranger than those of Weber in *Der Freischütz*. Far from being a simple process, the sophistication of harmony appears to ebb and flow like the tides.

Secondly, if it were indeed true that modern ears no longer hear the 'older' discords as discords in fact, it would be impossible for us, or at least, totally unsatisfactory, to listen to any of the outmoded music produced with such primitive means. Yet we can listen to, and enjoy, Mozart or Brahms in a concert programme including also Bartók or Schoenberg, and we may well find that the points of tension in the earlier works come to us more intensely, and express a deeper pain, than in the new. Everything depends on the context. In a piece of plainsong, a tension as simple as an interval drop of a fourth is a real tension. In a motet by Byrd, a minor third is really heard as a dissonance. The context may vary even in a single work. In the last movement of Walton's Viola Concerto, there is a fortissimo *tutti* which clashes semitones together until it is almost physically painful to the ear, yet at the end of the work, a simple tritone, *pp*, between the soloist and a solo clarinet, carries overtones of unrest and grief which may remain with us longer than the earlier demonstrations. As we found in the flow of folk-song, a simple departure from the norm will create a tension in a simple context; more violence is necessary if the context is complex or already contains many dissonant notes.

The actual patterns of tension and resolution which can be created by harmonic and tonal means are the subject of most of the theoretical writing on music, and I do not intend to detail the various procedures in this present work. The shorter phrases are catalogued with great thoroughness in *The Language of Music* and the handling of harmonic progressions is described in textbooks of harmony. It is only necessary to say that complex patterns can be built up by using chords of more or less dissonance, interspersed with concords, and the effects of these patterns will be considered in later chapters.

Time Tensions

Tensions in the dimension of time can be created in music because human beings have, inborn in them, a sense of tempo and a sense of dividing up time into regular divisions of even duration. Any disturbance of these regularities acts as a tension. Our sense of tempo depends, basically, on the rhythms of the human body, to some of which we become accustomed even before birth—heartbeat, breathing, the rhythms of the brain, the swing of the arms and legs in walking, all give us some sense of the 'natural' tempo. Plato ascribes ρυθμός to όρκεσις, or bodily movement. To find the tempo which is natural to us, we can turn to the experimental psychologists. Vierordt[1] set various tempi with a metronome, and asked subjects to say which they considered 'fast', 'slow' and 'moderate' or neutral. 'Fast' corresponded on average to one tick every 0·42 seconds, or M.M. 143, 'neutral' to one tick every 0·64 sec., M.M. 94, and 'slow' to one tick every 1·07 sec., M.M. 56. Similar tests were carried out by Katz,[2] who found 'short' times were on the average 0·25–0·55 sec., 'comfortable' times 0·60–0·65 sec., and 'long' times over 0·65 sec. Benussi[3] was even more precise: 'very short' 0·09 to 0·23–0·25 sec., 'short' 0·24–0·25 to 0·58–0·63 sec., 'indifferent' 0·58–0·63 to 1·08–1·17 sec., 'long' 1·08–1·17 to 2·07 sec., and 'very long' over 2·07 sec. Using a summary of these figures and others provided by Pierre Fraisse[4] we can express these results in terms of metronome markings as follows:

Very slow	below M.M. 30
Slow	M.M. 30 to M.M. 50
Moderate, neutral	M.M. 50 to M.M. 95
Fast	M.M. 95 to M.M. 240
Very fast	over M.M. 240

The really interesting region is the neutral, 'comfortable' tempo, as this provides the norm by which we assess the others

[1] *Der Zeitsinn nach Versuchen*, H. Laupp, Tübingen, 1868.
[2] *Z. Psychol. Physiol. Sinnesorg.*, vol. 42, Berlin, 1906, p. 302.
[3] *Arch. ges. Psychol.*, vol. 9, 1907, p. 384.
[4] *The Psychology of Time*, Eyre and Spottiswoode, London, 1964.

as fast or slow. This is the tempo between 50 and 95 beats per minute. There is another sense in which this is the neutral region in our ideas of duration: it is found in practical tests that people tend to show systematic errors in judging the duration of a sound, they overestimate short sounds and underestimate long ones. In between, there is a region where judgements are usually correct; this is known to psychologists as the *indifference zone*, and it has been found to lie in the region of 0·6–0·8 seconds, corresponding, for repeated sounds, to a metronome marking of 75 to 100.

It is instructive to compare the results of these psychological tests, conducted in the laboratory, with the experience of a cultivated and enquiring musician. Curt Sachs[1] writes: 'Men of today are generally unaware of the fact that there was, is, and must be, an average normal time—*tempo giusto*, as the time of Handel called it. Without the concept of normalcy we would not be able to rate a tempo as fast or slow. Again and again, each chapter of this book will have to record a standard time recognized as 'normal'; and again and again, each chapter of this book will have to mention that the regular stride of a man walking leisurely has provided the physiological basis. This implies a rather consistent time unit or 'beat' of 76–80 M.M....'

Sachs suggests that the maximum of slowness, still allowing for a steady step or beat, is possibly M.M. 32, but probably higher, and the maximum of speed, 'beyond which the conductor would fidget rather than beat', is probably M.M. 132. (Of course, a composer may *mark* slower or faster tempi, but these will tend to be modified by the performer or conductor and read as, in the slower case, two or more metronome ticks to each note. A composer may actually instruct the player to adopt, say, $\downarrow. = 15$, but in fact the player will set his metronome so that $\downarrow. = 30$ or $\downarrow = 45$. Similarly, if the composer sets an absurdly fast rate of ticking for the metronome, say $\downarrow = 200$, it will in practice be set to $\downarrow = 100$.) Sachs describes, as a personal experiment, trying to metronomize Bach's B minor Mass, where

[1] *Rhythm and Tempo*, Dent, London, 1953.

he found that the average of tempi over several movements and several days came eventually to a beat of M.M. 80. Thus Sachs's standards for fast, *tempo giusto*, and slow, fall very neatly into the ranges arrived at by the experimental psychologists such as Fraisse (who were not in fact concerned with music at all). As a further experiment, the reader may care to imagine performances of slow, fast and medium tempo music, and then check the actual metronome tempo: as an example of very slow music, I found 'Saturn' of Holst's *Planets* to have an M.M. rate of approximately 34 in the first section, where the slow swinging chords on the woodwind tick away eternity, and the frenetic 'Danse Générale' of *Daphnis and Chlöe* to be M.M. 190 (if one counts each of the five beats in the bar): these figures seem to bear out the previous assessments fairly well. In general it seems that we accept a tempo around M.M. 80 as normal, and tempi twice as fast or twice as slow approach the limits of our normal sense of time.

Sachs suggests that the rhythm of the normal stride in walking gives us our sense of tempo, and this idea is certainly plausible. However, it is not entirely facetious to suggest that, if this were the case, the length of one's legs would make a profound difference to one's sense of time, and there is no evidence that short-legged conductors habitually set faster tempi than lanky ones. A more likely hypothesis is that the speed of walking which we adopt is itself controlled by some other 'internal clock' which also affects our ideas of musical tempo. There is a great deal of evidence that this is in fact so, and that there is a recondite internal metronome which controls a number of our rhythmic actions.

Wundt[1] found that 0·75 second, the duration of the indifference zone, and a time corresponding to M.M. 80 for repeated operations, was approximately the time necessary for the apperception of a number of 5 to 6 digits, and for the association of two words. He concluded from this and other experiments that 0·75 seconds is approximately the time during which

[1] *Eléments de Psychologie Physiologique*, trans. Rouvier, Alcan, Paris, 1886.

the process of association is most easily accomplished, i.e. the time necessary for a word to be communicated by the ear to the brain, identified, compared with other words in the memory store, associated with another selected word, and for the selected association to be constructed by the vocal organs. The controlling process, in terms of speed, in all this is the mechanism for comparing the first word with a long string of remembered words which are potential associations: in short terms, 0·75 second is the minimum time for identifying a concept and then acting on it.

Fraisse (loc. cit.) has collected more modern evidence for the psychological significance of 0·75 seconds as a rate-controlling constant of the brain's mechanism. 'The duration of 0·75 second seems to be a psychic constant corresponding to the duration of the complete process of perception. . . . As regards behaviour, the efficacy of a sensation appears to reach a maximum when it precedes the reaction by about 0·75 second . . . if the conditioned stimulus (in a conditioned reflex) consists of a combination of two factors, conditioning is most easily established when there is an interval of about 1 second between them . . . subjects who are asked to reproduce a sound stimulus by pressing a button begin to react about 0·7 second after the stimulus has ceased, as if this interval were the optimum for immediate succession.'[1] He quotes many other researches in which the same interval of time emerges as the average time needed to perceive a situation and act on it.

The connection of this natural tempo with walking pace and heartbeat is not a coincidence. Though we are not conscious of its activities, the brain is controlling all our muscular processes, and it seems natural and economical that for the rhythmic ones, it should select a tempo that is already in use, so to speak. Fraisse comments: 'Walking, heartbeats, movements effected at a spontaneous tempo, and perceptions all follow on at intervals of about 0·7 second, which we consider to be the optimum interval for the functioning of the nervous centres because it is the most economical.'

[1] Oléron, *Année Psychol.*, vol. 52, Paris, 1952, p. 383.

This synchronization between various rhythmic activities is, indeed, one of the characteristic behaviour-patterns of the nervous system, both in man and the other animals. Sherrington showed that the scratch-reflex in the dog tended to set up a number of similar and synchronized rhythmic movements in its body, apparently controlled by a centre, and independent of the rhythm of any external stimulus. Similarly the sinus node in the human heart appears to act as a 'pacemaker' for a large number of other rhythmic internal functions.

Having shown that there is a *tempo giusto* of physiological origin, it is as well to consider how far the actual preferred tempo may vary from one individual to another, as this may go far to explain disagreements about the tempi to be chosen for various musical works. Obviously such differences may arise from physiological or mental causes.

Physiologically, all bodily functions are affected by temperature, normally becoming faster as the temperature rises. For rhythmic activities the alterations can be expressed by Van't Hoff's law in the form $\log f = c - (\mu/2\cdot3\ RT)$, where f is the frequency of the process, c and μ constants, R the Gas Constant, and T the absolute temperature (° Kelvin). For every type of reaction μ assumes a constant value, known as the *temperature characteristic*, approximately 29,000 for the human heartbeat, for instance. Fortunately the thermostatic activity of the human body keeps us at a fairly constant temperature and saves us from the fate of cold-blooded animals that rush around in the noonday heat and become dangerously slow in their reactions as dusk falls and their bodies cool. The weather does not affect our sense of tempo. However, if the body temperature does change, our internal clocks speed up with rising temperature and slow down with a fall, and our sense of tempo varies accordingly. Some people spend most of their lives with body temperatures slightly higher or lower than the average (the composer Bernard van Dieren was reputedly one of the 'feverish' type) and we should expect to find real differences between their ideas of a 'moderate' tempo. Using the value $\mu = 24,000$

found for tempo judgements by Hoagland[1] it is easy to show that a 1° C. difference in body temperature will make a difference of about 12 metronome units in our assessment of moderate tempi.

Mentally, there is some evidence that judgements of tempo may be altered by conditioning. J. P. Foley[2] reports that, among 684 students undergoing various vocational training courses, 'preferred' tempi were faster among typewriting and power machine operators than for beauty culture, trade dressmaking, and domestic science students. C. W. Valentine[3] comments on this result that it may also be the result of preselection, in that 'those naturally disinclined for very rapid mechanical movements might have avoided such occupations as typing and power machine operating . . .'. This seems very reasonable as an interpretation.

In general, we have seen that, allowing for these possibilities of individual variation, there exists in the mind of most people a concept of *tempo giusto* at around 80–90 beats per minute, and anything noticeably faster or slower than this creates a tension. The tensions are of different types: intuitively one would say that slower tempi induce tensions of the 'in-coming' type (resignation, sadness) while faster tempi produce 'outgoing' emotions (excitement, exhilaration). The indefatigable experimental psychologists have not left it to intuition to have the last word. There is ample experimental evidence that this is in fact the case. M. G. Rigg[4] tested the reactions of 85 students to the same musical phrases, changed in tempo according to a random scheme, over the range M.M. 60–160. He found that increasing the tempo had a significant effect on the judgements made by the students on the character of the phrases. More judgements of 'pleasant' and 'happy' were recorded with faster tempi, while slower tempi produced more judgements of 'sad'

[1] *J. gen. Psychol.*, vol. 9, 1933, p. 267.
[2] *J. Social Psychol.*, vol. 12, 1940.
[3] *The Experimental Psychology of Beauty*, Methuen, London, 1962, p. 251.
[4] *J. exp. Psychol.*, vol. 27, 1940, p. 566.

or 'serious'. This was, of course, under conditions where the harmonic or melodic character of the phrases remained the same during all the tests, so tempo was the only variable. K. Hevner[1] carried out similar tests with whole compositions and obtained results of the same kind.

Sachs has shown convincing evidence in his extensive historical and geographical studies that the tempo around M.M. 80 is considered normal, neutral, neither fast not slow, in many different cultures, and has so been considered for a very long period of time. We can therefore say that musicologists, experimental psychologists and neurologists have reached striking agreement on this point from a number of different routes.

The other important human reaction to time is to break it up into comprehensible units or unit patterns, and this provides the origin of our sense of rhythm in music. If a series of identical notes, or any sounds, is carefully performed so that no one unit of the series varies from the others in pitch, duration, or intensity, our reaction to this is the strange one of, in effect, refusing to believe our ears. We insist on hearing the notes in groups, usually two, four or three, and usually believe that we 'hear' slight accents at the beginning of each group, although no such differences exist. Valentine (loc. cit.) has shown that this tendency, in practice, can make us ignore accents which really *are* there: he tapped out an irregular rhythm, such as:

♫ | ♫♫ | ♫♫ | ♫ | ♫♫♫ | ♫

and so on, slightly accenting the beginning of each 'bar', and found that the great majority of his subjects, including music students, 'heard' the series as regular twos, fours or threes.

The general consensus of results in such tests is that groups of two are most commonly read into a series of notes, then groups of four. Groups of three seem less 'natural', and groups of five or seven are hardly ever reported. This certainly accords with the great mass of Western European music, 'square' rhythms of

[1] *Amer. J. Psychol.*, vol. 49, 1937.

twos or fours being very common, three rather less common, and other units relatively rare. The ecclesiastical convention of calling triple time 'perfect' and duple time 'imperfect' in earlier centuries is hardly evidence for the reactions of the normal populace. As with harmony, there was a tendency in the Church, almost a conscious one, to reject for ecclesiastical use everything that smacked too strongly of the secular music or the tastes of the common people. There is a sense in which the 'pure' fourths and fifths of *organum* were preserved, not for their musical value, but because they represented a reaction against the vulgar thirds and sixths of the secular tunes: the same reasoning applied, with more force, to rhythm.

In the East, of course, there has not up to now been the same interest in such simple rhythms and irregular time units are common in classical Indian and Chinese music. However, here again it is significant that 'square' rhythms occur in the folk music of these countries, and have also been readily adopted from Western music wherever there has been any contact with the West. There seems, therefore, to be a universal tendency towards simple time units of two, four and three steady beats.

Any departure from the normal pulse of the music then comes as a point of tension. This may occur as a change of time-signature, the use of an unusual unit, such as five, seven or eleven beats in the bar (the last particularly characteristic of Russian music), syncopation, where the expected accent is displaced, and various types of cross-rhythm effect, such as hemiola, or the alternation of six beats in the form 3 + 3 and 2 + 2 + 2.

Here again there is an element of relativity and a definite law of diminishing returns. In a square four-beats-to-the-bar sequence a simple syncopation, or an occasional two-beat bar, can create a strong tension effect which is resolved by the reversion to the normal rhythm. If every bar and every part is syncopated we soon begin to lose any real sense of tension and resolution, and there only remains a sense of ill-defined unrest. The rhythmic complexity of a good jazz group, for instance, is only effective against the steady four-beat background of the

38

rhythm section: the same parts without the background would merely be chaotic and irritating. Very often it is possible to create complex rhythmic effects within a simple time-signature. In fact, there is every likelihood that effects so created will be aesthetically satisfactory, because the composer has obviously been able to see the relationships between the tension-producing device and the basic pulse against which it is a tension. There seems to be a vulgar notion that rhythmic subtlety is only expressed in complex or variable time signatures. Dallapiccola has remarked, for instance, that Wagner's lack of rhythmic sense is shown by the fact that 'there is only one five-four bar in the whole of his output', which is rather like saying that Michelangelo's lack of a three-dimensional plastic sense is proved by the fact that he never carved holes in his statues. Whether Wagner had a strong rhythmic sense or not, the criterion seems rather silly. Brahms, who was almost equally conservative about his use of time signatures, in fact uses all the most effective rhythmic devices, and in many cases produces a more interesting feeling of rhythmic variety than Stravinsky at his most fidgety, with different time signatures for nearly every bar. Curt Sachs has pointed out, indeed, how often the time signatures in Stravinsky's music are quite arbitrary: the only useful function of dividing music up by bar-lines is to mark the positions of the most important accents at the first beat of each bar. This applies whether the bars are in constant or variable times. Stravinsky tends to divide his music into bars of irregular length, and then often adds *sforzando* marks on other beats than the first, which thereupon become the most important accents. This destroys any value which the barring might have as a guide to performance. Stravinsky's most interesting rhythmic effects, in fact, are in the passages where he *destroys* the sense of rhythm and reduces the music to a long series of notes of equal duration, following the number of syllables in the lines of the text, for instance, as in *Les Noces*. Compared with the complex but springy rhythms used by Bartók in, say, the scherzo of the Fifth Quartet, Stravinsky's devices are extremely crude.

39

The tensions of duple and triple time tend to be of different kinds. Duple time tends to be very regular, rather rigid, and stronger than triple time, which has more of a swing or lilt to it. The effects of alterations to the basic pulse are therefore also different: they are more disruptive to duple time. On the other hand, when these rhythms are considered in conjunction with tempo, the triple beat at high speed tends to become more abandoned and frenetic than the double beat. Deryck Cooke (loc. cit.) points out the 'comparatively controlled gaiety of the duple-time scherzos of Brahms's Fourth, Tchaikovsky's Fourth or Borodin's Second Symphony, with the abandoned excitement of the triple-time scherzos of Beethoven's Seventh and Walton's First'.

It is interesting to note that Hevner[1] has found similar results in an analysis of the psychological effects of poetry. She found that the two-syllable foot tends to be solemn, earnest, dignified and sad, while the three-syllable foot is joyous, light, sparkling, spirited and gay.

Volume Tensions

It is almost unnecessary to state that dynamic tensions can be set up by making part of a passage louder, or by building up a crescendo. This seems a 'natural' language of musical gesture, with analogies to speech, and it is only apposite to say that dynamic patterns, like any others, depend on the general level of volume for their effects.

Dynamic patterns on the small time-scale are really an aspect of rhythm, as they tend to mark the accented notes in a sequence, and volume really comes into its own as a pattern-making activity only in the larger-scale aspects of a work. For instance, the prevailing mood of Tchaikovsky's Sixth Symphony could not be adequately expressed merely by the pitch and time tensions, without the sudden bursts of sound, the sudden changes from piano to fortissimo, and the quiet brooding of the opening eventually fading away into inaudibility at the despairing close.

[1] *Amer. J. Psych.*, vol. 49, New York, 1937.

The Romantic movement brought a great change to the musicians' idea of dynamics—even in solo works one has only to think of the storm of sound which forms, say, the middle section of Chopin's Ballade in F to realize the urge towards dramatization by dynamic means, and in the orchestra the change was revolutionary. The modern reaction against such large forces may be in part a reaction against what seems almost too easy a method of playing on the hearer's emotions—it may also be a recognition of the law of diminishing returns which we have encountered in other aspects of musical language.

It is interesting to find, almost at the watershed between the old and the new attitudes, Holst's personal reaction against the enormous orchestra of the *Planets Suite*, leading to the modest scoring and unsensational dynamics of the Fugal Concerto and the other later works. He was obviously well aware of the dangers of overindulgence in dramatic noise. The remarkable passage in 'Uranus' of the *Planets* where a *ffff* chord of C major suddenly collapses to a *pp* dominant ninth on a handful of strings is meant to suggest the astrological character of Uranus, who is inclined to lead the individual into situations encouraging overwhelming *hubris* and then remove all his support suddenly, precipitating an equally overwhelming disaster. The lesson for the late Romantics needed to be learned.

Timbre and Texture

These elements in the total tension/resolution pattern may be discussed briefly, though their use is almost as obvious and natural as the use of dynamic changes. It is a natural extension of the emphasis afforded by playing a note or passage louder to perform it on a more penetrating or individual-toned instrument. The entry of the trombones in *Don Giovanni* is a famous example where no other musical device could have achieved the dramatic effect so economically. In many ways the chill of the 'supernatural' trombones marks the end of the work as a comic opera and the beginning of the great drama with which it ends.

Schoenberg, in his *Harmonielehre*, looks forward at the end of the book to a kind of 'melody' which strings together different tone colours as our present melody strings together notes of different pitch. This is obviously important for an art such as dodecatonalism which has the declared policy of abolishing the tension/resolution patterns of harmony: where one of the great pattern-making techniques is deliberately rejected, it becomes necessary to exploit the others as fully as possible. However, here again it is important to establish some kind of norm against which different tone-colours or textures appear as tensions. If the music drifts constantly from one colour to another, all sense of tension is lost.

The larger-scale units of pattern which can be created with those tension elements we have discussed form so large a subject that it would require a textbook of harmony, rhythm, form and orchestration to do the matter justice. We have discussed the tensions produced by tonal devices, whether intervallic or harmonic, by variations on the *tempo giusto* of about 80 beats per minute, by alterations in the steady metre of duple or triple time, by dynamic changes, and the modifications possible with timbre and texture. Because, for convenience, these have been considered separately, there may remain an impression that they act separately in conveying musical thought to the listener: this is not so. While in any particular work or part of a work we may feel that the harmonic elements or the rhythmic elements are playing the major role, the total work of art is a combination, an interaction, of tension/resolution patterns in all the modes available to the composer. This adds to the subtlety and complexity of musical thought, as at one time all the elements may be producing parallel patterns (as, for example, at the end of a Beethoven symphony, when harmony, tempo, rhythm, dynamics, colour and texture all convey the thought of a triumphant finale), or they may even produce the subtle shadings when, for example, a triumphant phrase (from the harmonic and melodic point of view) is echoed at a slow tempo and with low volume, giving an effect of remembered glories rather than present ones. (The distant military band in Elgar's *Cockaigne* Overture is a fine example of this effect.) It is, in reality, the interactions of all the musical elements which make music the flexible language which it is. In the next chapter we will consider the channels through which this language reaches the brain and how it can be interpreted in human terms.

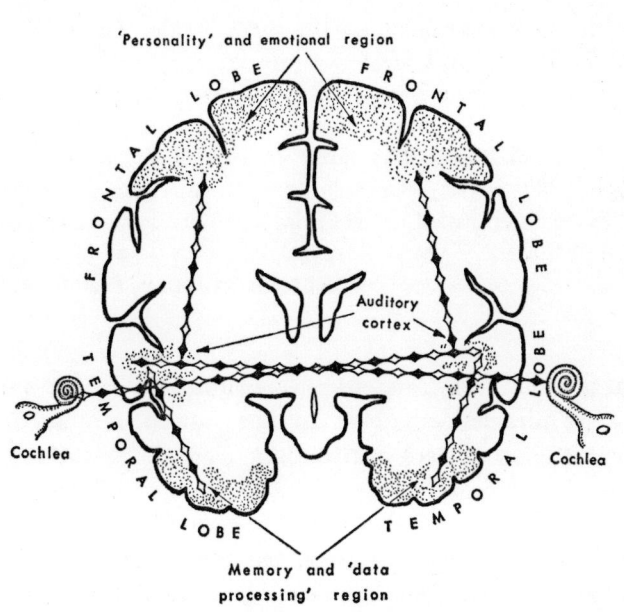

'Personality' and emotional region

Auditory cortex

Cochlea

Cochlea

Memory and 'data processing' region

(◆-◇-◆-◇ Diagrammatic paths of nerve impulses)

SIMPLIFIED VERTICAL CROSS-SECTION OF THE BRAIN

When the cochleas of the ears are stimulated by sound, patterns of electrical pulses pass from them to the two regions of auditory cortex in the brain, the signals crossing on the way. From the auditory cortex the patterns are passed to the data-processing regions in the lower (temporal) lobes of the brain, and to the emotional regions in the upper (frontal) lobes, for analysis.

3 Music and the Nervous System

Pongileoni's blowing and the scraping of the anonymous fiddlers had shaken the air in the great hall, had set the glass of the windows looking on to it vibrating; and this in turn had shaken the air in Lord Edward's apartment on the further side. The shaking air rattled Lord Edward's *membrana tympani*; the interlocked *malleus, incus* and stirrup bones were set in motion so as to agitate the membrane of the oval window and raise an infinitesimal storm in the fluid of the labyrinth. The hairy endings of the auditory nerve shuddered like weeds in a rough sea; a vast number of obscure miracles were performed in the brain, and Lord Edward ecstatically whispered 'Bach!'

ALDOUS HUXLEY, *Point Counterpoint*

In 1923, when *Point Counterpoint* was published, 'a vast number of obscure miracles' was about all that could be said about that particular series of steps in the chain, from the agitation of the ends of the auditory nerve to the recognition of Bach in the brain. The anatomical detail had come basically from Helmholtz in the nineteenth century, but the anatomist could not follow the musical message when it entered the mysterious world of the nervous system. It was known that nerves were somehow electrical in their action, yet they could produce chemical effects at their ends, and that was all.

Modern techniques, and particularly the rise of the electronics industry, have given the scientist some of the tools he needs to begin to trace the path followed by the musical thought when it leaves the *cochlea* or labyrinth, and the information which has been collected up to now sheds some very interesting light on the behaviour of the brain when such a message is fed into it,

44

behaviour which can give us a clue to the nature of the effects which music produces.

The sounds which become vibrations in the fluid of the cochlea set the appropriate hairs on the nerve endings vibrating in sympathy. Helmholtz had shown how the cochlea behaves rather like a microscopic harp, in that different regions of this tiny organ react in sympathy to notes of different pitches. If one of the hairs moves, it distorts the body of the nerve-cell or *neuron* to which it is attached, and this distortion marks a new phase in the existence of the musical message. Up to this point, whether as a vibration on a string, or in the air, or in bone, or in fluid, or along a microscopic hair, the sound has been represented by a physical *movement*. The movement goes as far as the neuron, and then this cell changes the message into a new form, an electrical potential.

The interior of any neuron is a tiny droplet of fluid, which is normally maintained at a voltage of -70 millivolts compared to the outside of the cell. This maintenance of a voltage difference between the inside and outside of the cell wall is characteristic of all cells, whatever their function and wherever they occur— nerves, muscles, skin, in plants, in bacteria, in animals—every unit of living matter with a cell wall has a difference of electrical state on the two sides of the wall; this is fundamental to the life and nutrition of the cell.

The effect of distorting the cell wall of one of the neurons connected to the ear is to break down this electrical difference for a fraction of a second. This momentarily changes the voltage of the body of the cell, so that its potential difference, compared to the outside world, is now no longer as much as -70 mV. The change of voltage spreads through the body of the cell and arrives at the base of the *axon*, which is a long tentacle-like growth from the neuron body. These long axons are features which characterize nerve-cells: some of them are 100,000 times as long as the diameter of the tiny cell body, and they act essentially as the 'wires' carrying messages from nerve-cell to nerve-cell and eventually to the brain.

If, as a result of the distortion, the electrical potential at the base of the axon drops to about −6omV, the axon begins to change its own electrical character. Sodium ions held outside the cell membrane begin to flow in, and a smaller number of potassium ions flow out: the total effect is to introduce a large number of positively-charged ions and the electrical potential at the base of the axon collapses from −70 mV to a small *positive* charge. This alteration in potential affects the next length of axon, so that this in its turn becomes electrically positive, and so the effect spreads to the end of the axon; it has been well described by a neurophysiologist as a 'kind of chemical smoke-ring'. By the time the electrical disturbance has reached the end of the axon, the cell body and the base of the axon have returned to normal and recovered their negative voltage. The total effect is that a mechanical disturbance to the cell wall has been converted into an electrical pulse which has passed all the way up the axon to the end.

There are three main advantages which arise from this seemingly complicated method of conveying an electrical message. The first one is that there is no loss of power. As has been said, axons are often extremely long compared with the size of the cell body. If the tiny electrical pulse, only one-hundredth of a volt, had to travel this distance by normal conduction, it would be completely eliminated by resistance in the axon before it has travelled a fraction of an inch. In practice, the pulse system enables the signal to pass along axons which are over a foot long—if someone treads on my big toe, for instance, the message reaches my brain, six feet away, over a chain of only three or four axons. It is rather like the difference between delivering a message over twenty-six miles by one marathon runner or a hundred relay runners: the marathon runner arrives completely exhausted, while the relay runners have only to run about a quarter-mile each, and the last man will arrive very little more tired than when he set out.

Secondly, the system guarantees one-way traffic flow. Messages can only pass from the cell-body to the end of the axon, and not the other way. This system saves us from the risk of a signal turning back halfway up the axon, or going to some other organ when it was intended for the brain.

Thirdly, the system has a degree of resistance to chance signals. The signal does not start unless there is a difference in potential from normal of about −10 mV, and therefore there is a 'threshold' below which very small disturbances do not have any effect. This saves us from the state of being so 'over-sensitive' that we react to any change in our surroundings, no matter how minute or unimportant, and it also protects us from reacting to accidental electrical impulses set up by some fault in one of the millions of nerve-cells in our bodies—what the communications engineer would call 'noise in the circuit'.

The Synapse

When one of the electrical pulses reaches the end of an axon, it will in every case be near another neuron, either in a bundle of nerve fibres or in the brain. There is a microscopic gap between the end of the axon and the next cell, and this gap is called a *synapse*. These tiny straits, about one-millionth of an inch across, have a profound influence on the whole system of communication in the body.

At the synapse, the electrical impulse which has been blocked in its further progress by the gap at the end of the axon releases a small quantity of a transmitter substance into the fluid that bathes the gap between the cells. This substance migrates to the cell wall of the adjacent neuron and modifies the permeability of this membrane so as to allow a slight positive charge into the body of the cell (this is, indeed, much the same effect as is achieved by mechanical distortion of the cell, as we have seen earlier in the process). The potential of the second cell falls, and, as before, if it falls below −60 mV the affected cell will send an impulse along its own axon, thus passing on the message to the next cell in the series, and so on to the brain. Most cells have several axons touching them through synapses—a point which may have been obscured because we have dealt

47

with single cells for simplicity. This fails to convey the complexity of the system and the extremely large number of neurons involved in even the simplest operation. There are about ten thousand million of them in the human body, and it is likely that on most occasions when sense information is being passed to the brain, two or more of them are sending out electrical pulses simultaneously to the neuron further up the chain leading to the brain. If the threshold for activity—the necessary voltage change for the message to be passed on—is set higher than the potential change caused by *one* pulse reaching the cell wall through a synapse, but lower than the potential change resulting from *two* pulses, the third cell will not 'pass on the message' until it has received it from two other cells almost simultaneously. This is another precaution against us reacting to a chance electrical disturbance in the nervous system—it is unlikely that two neighbouring cells will go wrong at exactly the same time.

An interesting property of the synaptic gap, which has great importance in the shaping of learning processes and memory, is that it becomes easier to pass the gap the more it is used. This idea was first put forward as a hypothesis to explain the retention of memories, by the brilliant neurologist and neurophysiologist, W. Ritchie Russell,[1] and evidence has since been building up to support his point of view with experimental results. I shall return to this point when we consider the function of memory as a brain operation, but it is useful at this stage, while considering individual cells, to note that Russell's theory supposes that the electrochemical 'transmitter substance' has a long-term effect in lowering the effective resistance of the synapse to any subsequent pulses. This explains such everyday experiences as learning a piano piece, a golf swing, or a poem, by repetition; we should expect the repetition to 'open' circuits so that the fingering, the handling of the club, or the words of the poem, will 'come back to us' readily with a small amount of appropriate stimulation. It also explains the fact, familiar to all doctors dealing with concussion cases, that old memories are far

[1] *Brain, Memory, Learning,* University Press, Oxford, 1959.

harder to remove from the brain than new ones—for instance, the events leading up to an accident very often fade away completely from the mind of the injured person, while the memories of days, months, or years before are usually easily revived. This would obviously occur if the old memories had been confirmed in their neuronal pathways by the easing of the synapses concerned. Russell takes the idea a step forward. As 'noise in the circuit' is constantly being produced by the random firing of neurons (hence all the precautionary devices against us acting on such accidental signals), it would be expected that random signals in such a 'memory' circuit would continually traverse the easy synapses, and in doing so, make them even less resistant to future electrical activity.

The Electrical Pulse

It is clear from the method adopted by the nervous system for passing its messages that the electrical pulse which reaches a synapse has had no physical contact with the exciting cause of the signal. It is the last runner in a relay. It is equally clear that the pulses must all be the same, because they are controlled only by the chemical and electrical nature of the neuron, and are not affected by the outside stimulus except in an 'all-or-nothing' sense—a stimulus triggers off a sequence of pulses: after that it can have no further effect. Yet it is a fact of experience that we can somehow tell the difference between a strong sense impression and a weak one—a loud noise or a quiet one, a sharp pain or a mild one, a bright or dim light. There must be a mechanism for differentiating these impressions, using a code composed only of identical electrochemical pulses. (The evidence for the fact that the pulses *are* identical is too lengthy to review in this chapter: it commenced with a series of crucial experiments by F. Gotch at Oxford in the early years of this century, and an enormous body of evidence has since been built up that the pulses do not vary in intensity of electrochemical activity, or in any other way, and that the pulses are essentially identical in every nerve cell.)

D   49

This differentiation is provided, not by any variation in the strength of the pulse, but by the frequency with which they are produced. Thus a very quiet noise may provoke only a few pulses per second in one of the nerves connected to the cochlea, while a loud one will produce several hundred per second. The actual pulse which passes shows a rise to a peak (positive) voltage followed by a slower decline to the normal negative (−70 mV) level. The whole operation takes about one-thousandth of a second, with the peak voltage lasting about one-tenth of this time. As the pulse passes up the axon, the base returns to its normal voltage and is therefore ready to fire again. The frequency of the pulses increases with greater intensity of the stimulus, until a point is reached at which no further increase of frequency is obtained no matter how intense the stimulus. This is when the frequency is approaching one thousand pulses per second, or, in fact, when a new pulse is being propagated as soon as the previous one has left.

At the other end of the scale, the very slow frequencies may not come to a synapse often enough to activate the next cell in the chain—they do not achieve the right potential to reach the threshold of the synaptic gap—and therefore the signal goes no further. This accounts for the threshold of hearing; similar threshold effects are experienced with every sense—touch, sight, smell and so on.

As has been said, the pulses are identical whatever the intensity of the stimulus, and they are identical in every nerve, whether from the ear, the eye, a chemical receptor in the nose, a Pacinian corpuscle for measuring pressure on the skin, or a touch or cold receptor. The same code is used not only for signals going towards the brain, but for signals coming from it to the muscles and glands, along the *effector neurons*. The only indication of their message is the organ from or to which they travel and the route which they take.

The Brain

When the signals reach the brain, it is rather like the country boy from a tiny village coming to the great city. The few million axon paths which constitute the spinal cord run into the thousands of millions of neurons forming the grey matter of the brain, with its 300-odd million axon paths and connections forming the white matter below. It is not necessary, or practicable, to go into the detailed anatomy of the brain here, but it is worth recording that it consists of three main parts, the *cerebellum*, or small brain, situated at the base and rear of the skull, the *brainstem*, which is really an extension of the nervous connections of the spinal cord, and the two hemispheres of the *cerebral cortex*, a six-inch thick sheet of grey neuron cell-bodies, folded and convoluted so as to give as much effective area as possible, and surmounting the white mass of axons.

Faced with this enormous mass of connected and interwoven neurons, physiologists have been forced to evolve techniques for mapping the geography of the brain. This is relatively easy for the motor areas, those which control the movements of muscles, as an artificial electrical pulse can be sent from this region, and the physiologist can see which part of the body moves. An electrical probe embedded in the brain (the brain itself contains no pain receptors, and the insertion and maintenance of a probe involves very little discomfort) at various points, and activated with an appropriately tiny electrical charge, will simulate real effector activity, the pulses passing from the probe and making the corresponding muscles work. For instance, a probe inserted into various points in a strip of grey matter running across the cerebral cortex, just forward of the middle of the skull (the *precentral gyrus*) will cause involuntary movements of the toes, if the probe is well down in the *great longitudinal fissure* which divides the hemispheres, while it will move the leg at the hip if it is near the edge of the fissure. The same sort of pulse will cause a twitch of the eyelid if the probe is halfway down the side of the precentral gyrus, or the lips, if it is lower down.

It is interesting to note that the extent of the motor control area does not depend on the size of the organ it controls; it depends on the complexity of the movements which are required. For example, the hand is immensely more complicated in its repertoire of movements than the leg, and it has, consequently, a much larger area of control in the brain. The size of the operation is immaterial: a simple signal to my *rectus femoris* in the thigh will make it lift several pounds of leg: a comparable set of signals sent to the lumbrical muscle of my index finger may serve only to form the tail of a letter as I write. Pigs have a similarly large working area of the brain for their snouts: as E.D. Adrian says, on the subject of the practical difficulties of mapping the brain—'Here the whole of the tactile receiving area seems to be devoted to the snout, and if messages from the limbs reach the cortex at all it must be in some small area where they have so far escaped recognition. The pig's snout is its chief executive as well as its chief tactile organ, spade as well as hand, whereas the legs are little more than props for the body.' This passage comes from Adrian's Waynflete Lectures, 1946,[1] and this book remains one of the classics of the subject both for its clarity and for the elegance of the language in which this great physiologist reviews his special subject.

Behind the motor area there extends the *postcentral gyrus*, a parallel area dealing with the messages sent from the organs controlled from the precentral gyrus. As, in general, complex organs like the hand, eyes, mouth and so on, need to be richly endowed with sensory nerves, so that we can tell exactly what all the muscles are doing, the areas of the brain concerned with receiving these messages are roughly in the same proportions as in the motor areas. The lips and tongue, for instance, are endowed with an enormous number of heat, cold, touch, pressure and taste receptors, an apparatus designed to help us find suitable food, but, because of its rich innervation, capable of other pleasant uses.

The special senses, as might be imagined, send their messages to special parts of the brain. Messages from the retina of the eye travel through to the *occipital region* at the rear of the

[1] *The Physical Background of Perception*, University Press, Oxford, 1947.

brain: the olfactory nerves end in the frontal region and the auditory nerves run to areas on the side of the cortex—left ear to right hemisphere and vice versa. Application of electrical probes to the visual cortex will give the hallucination of flashes of light: application to the auditory area will produce the impression of a series of clicks or buzzes.

This mapping of the brain (for further details see Adrian (loc. cit.) and A. R. Buchanan[1] for more modern information in this rapidly-developing subject) only brings us to the fringes of our quest for the point where sense stimuli become music. At the area of the auditory cortex, all that we can say is that something is going on in the ear: we have no consciousness of it and no cognisance of its nature. There must be a 'data-processing' stage for us to identify the type of activity which is coming to us through the ear.

The receptor nerves have provided information in the form of a signal. This has to be compared with memories of similar events which have happened previously, its character assessed by these comparisons, a decision taken about the appropriate action, and orders must be transmitted to the corresponding muscles or glands if the action is a positive one. Some of the simpler processes hardly get to the brain proper; they are controlled by *reflex arcs*. Such a reflex arc gives us the useful capacity to close our eyelids rapidly if we see anything threatening the eye. The speed of the decision which operates the *orbicularis palpebrarum* (which muscles close the eyes) is very much greater than the data-processing operation which is necessary to identify the object—to decide whether it is a tennis ball or merely a shadow. Our eyes have already closed by the time we know what the threat really is. Similarly, if we tread on a tack, we jump away before we feel the pain: the reflex arc removes us from the dangerous object through a chain of nerves mainly in the lower part of the body and these have already done their work by the time we feel hurt, which is a cortical operation.

[1] *Functional Neuro-Anatomy*, Lea and Febiger, Philadelphia, 1961.

Many lower animals depend entirely on these simple chains of neurons for their whole existence. For instance, the sea urchin turns its pointed needles to face an approaching enemy—behaviour that we might think as 'intelligent'. However, this response is completely automatic and localized. A small chip of a sea urchin's spiny coat, carrying only one spine, will still turn its single defence against an enemy. It appears that the coat has chemical receptors which 'taste' the water for traces of organic matter from other living creatures. Human behaviour employs many thousands of such reflex arcs, and many of them are by definition unconscious. We decide consciously to raise our arms to stop a bus: the brain organizes the surprisingly complex series of operations which stop us from overbalancing as the weight of our arm distorts our weight-distribution. We could not do without reflex arcs to keep us out of trouble, but in the appreciation of music, or any conscious behaviour, they do not contribute.

The data-processing stages—comparing each incoming experience with previous experiences, looking for similarities, sorting out those signals on which we want to concentrate, volition, self-consciousness—are all outside the scope of the simple mechanisms we have considered, and, unfortunately for the physiologist, they cannot be analysed by such simple techniques as sticking electrodes into the brain. In the early stages, there was not even reliable evidence to say whether these activities were electrical in the same way as the nervous signals were electrical.

Brain Rhythms

The first evidence that these activities were electrical in the same sense as the reflex arcs came in 1924, when Hans Berger of Jena discovered that he could detect electrical events going on in the brain by means of metal strips pasted to the scalps of human subjects. His apparatus for detecting the very small voltages involved in these brain mechanisms (especially through the thickness of the skull) was not very accurate, and he had no means of amplifying the signals—the triode valve had in fact been invented by Lee Forest in 1907[1] but its use as an amplifier had not been sufficiently studied until the demand for radio receivers became important in the 1920s. In May 1934 Adrian and Matthews, using electronic apparatus, gave the first convincing demonstration of the 'Berger rhythms' or 'brain-waves' at Cambridge. From these small beginnings a whole new science, electroencephalography, or E.E.G. for short, has grown up, with several hundred laboratories and clinics devoted to its study, and a learned society publishing the latest findings in a journal.

The scope of this technological step cannot be over-emphasized. A great deal of experimental work had been carried out painstakingly on single nerve-cells isolated from the body—usually nerves of members of the cuttlefish family, such as the squid or octopus, which are useful for experiment because the axons are relatively large, easy to isolate and convenient to connect up to apparatus for measuring the electrical activity. A great deal was known about the properties of these thread-like cells before Berger ever applied electrodes to the human head.

At the other end of the scale, Pavlov and his collaborators had been studying the behaviour of complex animals such as dogs, but without any clue as to the exact mechanisms of the discoveries they made, such as the conditioned reflex. Berger's discovery opened a way to study the electrical activity of the living, active brain.

It must be said at once, however, that the move from isolated cells directly connected to instruments, to the measurement of the activities of the tens of millions of cells 'heard' through the thickness of the skull, was a move from a high order of accuracy and experimental control to a far lower order. To find out what is actually happening in the brain by E.E.G. is rather like trying to find out how a watch works by listening to the ticking. Inference has to be used a good deal and the science of psychology

[1] U.S. Patent 841,387.

has to reach down to meet the science of neurophysiology to see if the inferences make sense.

The first rhythms detected by Berger and later confirmed by Adrian came from the occiput, the site of the visual cortex, and it was soon discovered that these electrical pulses had a regular period of about ten per second. This electrical activity was christened the *alpha rhythm*: Berger and later workers were able to show that, apart from their origin somewhere around the site of the visual cortex, they had a deeper connection with vision. The rhythm was strongly marked when the eyes were shut, but diminished to a low level when the eyes were open or if the brain was kept otherwise busy—trying to remember a long string of numbers or performing mental arithmetic. The rhythms, though very weak compared with artificial electronic or electrical waves—for example, in a television set—are extremely important functions of the brain. The size of the electrical disturbance, which can be calculated from the external effects, corresponds to about a million neurons 'firing' simultaneously into their neighbouring cells. This is why the effects were so easily detected even with Berger's crude apparatus.

It was found very early in the study of E.E.G. that the alpha rhythm is noticeably affected by some types of mental disease, such as epilepsy. Very often mental abnormalities could be detected by the change in the alpha rhythm before the grosser symptoms of behaviour had become very much modified. Though the diagnosis was quite empirical, and no one knew, or knows as yet, precisely why the alpha rhythm is altered in this way, there grew up a large body of E.E.G. information of this type—vital to the psychiatrist, and constantly in use at every major hospital. Unfortunately it is not very informative about the basic workings of the brain. Just as a radiologist can tell you that a certain 'shadow' on the x-ray picture of the lungs portends tuberculosis, because he knows that such shadows occur with all tuberculous patients, or the physician can tell from the shape and colour of the spots whether you have heat-rash or smallpox, so the E.E.G. expert can detect an abnormal or injured brain from its electrical rhythms and actually help the surgeon by focussing on a particular point in the brain which is behaving oddly. This is particularly helpful in detecting brain tumours. However, these are purely descriptive or empirical sciences: ask the radiologist exactly what produces the shadow, ask the physician why smallpox produces spots and not an overall colour, ask the E.E.G. operator exactly what makes the odd rhythm, and they will rightly say that this is a matter for the research worker, not the active medical man.

However, the search for more refined methods of detecting abnormalities in the brain rhythms has encouraged the production of vastly improved, partially automated, apparatus which can also help the research worker. W. Grey Walter, who was instrumental in having built the 'Toposcope', one of these great advances in technique, recalls the difficulties of the early years:[1]'It was in the old Central Pathological Laboratory of the Maudsley Hospital in London, in 1929. The team there under Professor Golla were in some difficulty about electrical apparatus; they were trying to get some records of the "Berger rhythm" using amplifiers with an old galvonometer that fused every time they switched on the current. . . .' Such temperamental apparatus could hardly have become the everyday diagnostic guide which it is today, without the advances in instrumental design since 1929.

The improved apparatus has enabled the researcher in the field of the brain's electrical activity to recognize several different rhythms beside the prominent alpha rhythm—*delta rhythms* at 0·5–3·5 c.p.s., associated particularly with childhood and sleep, the *theta rhythms*, at 4–7 c.p.s., which mark the loss of pleasure ('whenever a pleasant situation comes to an end, a theta rhythm begins to appear regularly a few seconds later and culminates in about ten seconds, when it abruptly disappears...' —Grey Walter, loc. cit.), the *beta rhythm* at higher frequencies, 14–30 c.p.s. It is not my purpose to attempt to analyse these rhythms, but to indicate that they are tangible evidence that the activities of the brain are similar in kind to the electrical

[1] *The Living Brain*, Duckworth, London, 1953.

activities of the simple chains of neurons we have considered before, though vastly more complicated because of the enormous numbers of cells involved. Thought is achieved by the movement of pulses of electricity in the brain cells, just as sense-signals are carried by pulses of electricity in the nerves leading to the brain. The great neurophysiologist Sir Charles Sherrington described these electrical activities of the brain in a passage which has been much quoted, but which still seems so memorable that one despairs of finding better words—'A scheme of lines and nodal points, gathered together at one end into a great ravelled knot, the brain, and at the other trailing off into a sort of stalk, the spinal cord. Imagine activity in this shown by little points of light. Of these some stationary flash rhythmically, faster or slower. Others are travelling points streaming in serial lines at various speeds. The rhythmic stationary lights lie at the nodes. The nodes are both goals whither converge, and junctions whence diverge, the lines of travelling lights. Suppose we choose the hour of deep sleep. Then only in some sparse and out-of-the-way places are nodes flashing and trains of light points running. The great knotted headpiece lies for the most part quite dark. Occasionally at places in it lighted points flash or move but soon subside.

Should we continue to watch the scheme we should observe after a time an impressive change which suddenly accrues. In the great head end which had been mostly darkness spring up myriads of lights, as though activity from one of these local places suddenly spread far and wide. The great topmost sheet of the mass, where hardly a light had twinkled or moved becomes now a sparkling field of rhythmic flashing points with trains of travelling sparks hurrying hither and thither. It is as if the milky way entered upon some cosmic dance. Swiftly the head mass becomes an enchanted loom where millions of flashing shuttles weave a dissolving pattern, always a meaningful pattern though never an abiding one. The brain is waking and with it the mind is returning.'[1]

[1] *Man on his Nature*, University Press, Cambridge, 1940.

These dissolving patterns are thought—what may be, in our case, the appreciation of a musical pattern which has proceeded from the ear to the auditory cortex, and is now being processed. Lord Edward is analysing the sound pattern which has reached him from the great hall of Tantamount House, and is identifying it as Bach. Without processing—identification, comparison, judgement—the pattern is nothing to us. If we had no comparisons, no memory, no previous experience against which to compare the incoming pattern, it would be entirely meaningless.

No one knows exactly how the patterns are compared—whether they are sent from the auditory cortex (in our example —a musical pattern) to another part of the brain for checking against the 'files' of memories, or whether they are studied and assessed on their way to the auditory cortex from the ear. The fact that the brain has such a peculiar arrangement, with all the cortical reception areas arranged diametrically opposite to the sense organs which they serve, so that the brain is a mass of crossing lines, suggests that the great tangle in the middle may be designed to divert part of the signal pattern into the data-processing part of the brain, just as in a reflex camera the light can be diverted as it proceeds from the lens to the film, so that the eye can assess the picture. But this is hypothesis.

It is important to remember, whatever the mechanism of memory, that the patterns which are processed in the brain are patterns of electrical impulses, and are composed of exactly the same type of pulses as are used in all nervous activity. Therefore, as we have seen before, the only way that we know a particular pattern is 'hearing'—in fact, the only meaningful definition of hearing—is that these patterns are travelling from the ear to the auditory cortex. The same pattern of pulses may by chance be passing from the retina of the eye to the visual cortex, but in this case we will call the pattern 'sight'. Now this arrangement is very reliable as long as the signals stay in the straight neuron line between the ear and the auditory cortex, or the eye and the visual cortex. What do we make of them if they are diverted into a data-processing stage? There is nothing to show that pattern A came from the ear and pattern B from the eye, and if they happen, as we have postulated, to be identical apart from their source, they will *be* identical in any other part of the brain than the appropriate cortex areas. Grey Walter puts the matter clearly—'If, when you get a number on the telephone, you give a message, the message remains the same, even if you give it to a wrong number. The result of such an error in the brain is very different. Supposing some vinegar comes into contact with one of the sensitive end organs of taste in the tip of your tongue and 'gets a wrong number'—that is, say, supposing the nerve fibre conducting the impulse provoked by the vinegar, instead of connecting with its proper reception area, becomes in some way cut and grafted on to a nerve fibre leading from the ear to the brain—what do you think you would taste? You would taste nothing. You would hear a very loud and startling noise. Everytime the nerve end in your tongue was stimulated you would have a similar hallucination. If, instead, one auditory nerve were in this way misconnected with an optic nerve, when you heard music you would see visions. This is the mechanistic basis of hallucination. Such accidents are unlikely to occur in the peripheral nerves, where cross-talk is avoided by special anatomical arrangements, but they do occur in the brain itself.'[1]

One of the reasons why such confusions can occur in the brain itself is that the patterns are continually being 'run over' in repetition to compare them with older records of other patterns. Everyone must be familiar with the continuous and often cyclic programme of old associations and memories which can be brought into the open by some new experience—this is part of the data-processing activity of the brain, momentarily brought out by introspection. The 'easing' of the synapses, which we discussed earlier, helps to promote the setting up of such circulating memory stores. Sir Francis Galton, in a remarkable prevision both of modern neurological work and the psychological

[1] W. Grey Walter, loc. cit.

technique of word-association said, in 1883, 'I conclude from the proved number of faint and barely conscious thoughts, and from the proved iteration of them, that the mind is perpetually travelling over familiar ways without our memory retaining any impression of its excursions. Its footsteps are so light and fleeting that it is only by such experiments as I have described that we can learn anything about them. It is apparently always engaged in mumbling over its old stores. . . .'[1]

When something in these 'old stores' turns out to have some similarity to a new sense-impression, just arrived in the brain, we have the state of recognition. The two events do not have to be identical: the brain is adept at selecting the points in common possessed by the most dissimilar things. The data-processing behaviour of the brain, in fact, also involves the capacity to recognize the pattern underlying two events.

This brings us back to our two similar patterns, one from the eye and one from the ear. Unless the nerves have actually been grafted wrongly, as in Grey Walter's hypothetical situation, one of the ways in which we recognize the signals as sight and hearing is that, in addition to coinciding in the data-processing stage of the brain, they still reach the auditory and visual cortexes so that we *know* they are sight and hearing respectively. What becomes now of the identity which occurred in the other part of the brain? The answer is—an association. The pattern from the ear still remains 'hearing', but the fact that it can be identified with a similar pattern from the eye is not lost on the master pattern-analyser; every time we hear that pattern from the ear we shall be reminded, automatically, of the sight which produces a similar pattern. This is one of the brain processes called a hallucination: for those who are wary of this word there is a more exact one—synaesthesia.

[1] *Inquiries into Human Faculty*, Macmillan, London, 1883.

4 Synaesthesia

Grey Walter illustrated the fact that all sense impressions are processed in the brain using the same electrical code, by supposing a hypothetical misconnection in which the wires of the brain were crossed, and messages from one sense organ were confused with those from another. For obvious reasons such an effect cannot be brought about by surgery on human beings, and animals could not tell us what form the results took. However, examples of sense impressions becoming confused, or overlapping, because of natural effects in the brain, actually occur quite commonly without outside surgical interference. A message from the ear becomes confused with the visual pathways, so that the person involved actually sees colours or patterns when music is played: some people can 'feel' the texture of a picture or sound, and the sense of tactile quality is just as 'real' to them as the external stimulus. Signals from one sense organ, in fact, have shown so much similarity to those from another, in some way the brain recognizes that the two impressions become fused in the mind.

This widespread hallucination is called *synaesthesia*, and defined as 'a phenomenon in which a stimulus presented in one sense mode seems to call up imagery of another sense mode as readily as that of its own.'[1]

The word 'hallucination' must be treated with care. It suggests something rather unpleasant to many people, connected either with drugs, delirium tremens, or mental decay, or somehow connected with psychic powers, ectoplasm, auras and haunted houses. In fact, hallucinations are far more common

[1] M. D. Vernon, *Visual Perception*, University Press, Cambridge, 1937.

than is generally realized, mainly because most hallucinations are so mundane. Strictly, a hallucination is any mistake of the brain in its interpretation of the mass of sense-data which our eyes, ears, sense of touch and so on, are continually sending in to the brain. If I think I am falling physically as I fall asleep, one of the commonest mistakes, which must have happened to most people at some time or other, that is a hallucination. Peter McKellar[1] reports that, as an example, 144 out of 182 students recognized and had experienced this 'falling' sensation. It is so common, indeed, that possibly the phrase 'falling asleep' has some connection with it.

Francis Galton, that great nineteenth-century polymath, was one of the first investigators to make systematic researches into hallucinations, and synaesthesia in particular, and it was due to his patience, and his captivating degree of common sense in presenting the results, that we now know how common, and how commonplace, such experiences are, and how far the workings of 'normal' minds have curious by-ways.

By far the most common synaesthetic experience is to see colours or patterns when music is heard: to use McKellar's nomenclature, a *visual-auditory* synaesthesia. (The hallucinatory sense impression is put first, and the real stimulus which provoked it second.) This effect is also known as 'colour hearing'; it has been reported as far back as the time of Ptolemy. Sir Arthur Bliss, who is one of the many professional musicians subject to this trick of the brain, describes the experience as a constant play of colour sensations which pass through his mind when he is composing: the vividness of these effects must surely account for the extreme precision of the titles of the movements in his *Colour Symphony*:

(First movement):   Purple: the Colour of Amethysts, Pageantry, Royalty and Death.
(Second movement):   Red: the Colour of Rubies, Wine, Revelry, Furnaces, Courage and Magic.

[1] *Imagination and Thinking*, Cohen and West, London, 1957.

(Third movement): Blue: the Colour of Sapphires, Deep Water, Skies, Loyalty and Melancholy.

(Fourth movement): Green: the Colour of Emeralds, Hope, Joy, Youth, Spring and Victory.

Galton[1] quotes examples of associations with vowel sounds—'A is brown. I say it most dogmatically, and nothing will ever have the effect, I am sure, of making it appear otherwise. I can imagine no explanation of this association. . . . Shades of brown accompany to my mind the various degrees of openness in pronouncing A. I have never been destitute in all my conscious existence of a conviction that E is a clear, cold, light gray-blue. . . .' (Dr. James Key). For this particular person, I was black, O white and R copper coloured.

Rimbaud, in *Une Saison en Enfer* ('Délires, II') provides one of the best-known literary examples of this very personal type of visual-auditory synaesthesia:

'. . . J'inventai la couleur des voyelles!—*A* noir, *E* blanc, *I* rouge, *O* bleu, *U* vert—Je réglai la forme et le mouvement de chaque consonne, et avec des rythmes instinctifs, je me flattai d'inventer une verbe poétique accessible, un jour ou l'autre, à tous les sens. Je réservais la traduction.'

Isidore Isou [2]has taken the Rimbaud method to its logical conclusion by making up words with the vowels and consonants which suggest the right colour combinations, aiming at a pointilliste effect rather than verbal sense. Perhaps he feels that Rimbaud's translation rights have passed to him.

Obviously the 'crossed wires' which cause these involuntary associations between widely different experience do not have the same effects in every case. Dr. Key saw A as brown, E as grey-blue. Rimbaud saw them as black and white. Galton quotes enough examples to dispel any idea that there is a uniformity in this experience:

[1] *Inquiries into Human Faculty and its Development*, Macmillan, London, 1883.

[2] *Introduction à une Nouvelle Poésie et à une Nouvelle Musique*, Gallimard, Paris, 1947.

(Mrs. H.): '*A* pure white, and like china in texture;

> *E* red, not transparent; vermilion with China-white would express it;
>
> *I* light bright yellow; gamboge;
>
> *O* black but transparent; the colour of deep water seen through thick clear ice;
>
> *U* purple;
>
> *Y* a dingier colour than *I*.'

Mrs. H.'s daughters differed again in their associations: one saw '*A* blue, *E* white, *I* black, *O* whity-brownish, *U* opaque brown', the other saw *A* as black and *O* as white.

On the other hand, while there is not the faintest sign of consistency or agreement from one person to another in the associations produced by vowels or other sounds, they remain a very permanent and stable part of the individual's perception. As Galton says, 'They are not the capricious creations of the fancy of the moment, but are the regular concomitants of the words . . .' (he is writing here about word-colour associations) '. . . and have been so far back as the memory is able to recall.' Or as his subject, Mrs. Stones, says, '. . . I have always associated the same colours with the same letters, and no effort will change the colour of one letter, transferring it to another . . .' (in the diagram which she submitted to Galton for his statistical investigation, *A* is yellow, *E* green, *I* black, *O* red and *U* a kind of dove-colour).

It is important to stress the point of the reality of the synaesthetic sensations for the people who see them, and to establish that the images are not just associations called up at the end of a chain of conscious thought. As Galton, McKellar and other workers in this field have confirmed, the sense impressions which are evoked are as 'real' to the subject as actual colours from an external source, and they occur simultaneously with the sound or whatever other stimulus prompts them. There is no element of conscious thought in the process, and even if the synaesthetic impressions are unwanted or even ugly or frightening, there is no mental mechanism for preventing them. The reader with

E   65

no experience of synaesthesia or similar effects may find it difficult to understand, or even to credit, the reality of these phenomena to other people, much as a colour-blind man may well have his reservations about the 'reality' of colour vision, as described to him by others. It is salutary to remember that the enquiring mind can rise superior to such limitations: John Dalton, who was himself colour-blind (he gave his name to the condition, *Daltonism*), was the first investigator to discover and write about the differences between his vision and that of other people. Similarly Francis Galton, who had no personal experience of hallucinations, was the first person to analyse accounts of such experiences by other people, and to show that they were widespread and 'normal' elements of perception.

McKellar lists many other types of synaesthesia: such instances of identification seem to be possible between any of the senses—not only sight, hearing, touch, taste and smell, but the less commonly recognized human senses, such as body schema and kinaesthesia. Body schema is that mysterious sense which tells us how and where our bodies are disposed, whether we are upright or lying down, whether our arms are near our bodies or stretched out, and so on, without us having to look or feel consciously. Kinaesthesia is the sense of movement, for which the sense organs are mainly the semicircular canals of the ears. This sense tells us, in a darkened cabin on a boat, that the boat is swaying.

Simpson and McKellar[1] have listed the following combinations of sense impressions recorded as natural experiences:

Imagery	Sensory Stimulus
Visual	Auditory
Visual	Tactile
Visual	Gustatory
Tactile	Visual
Tactile	Auditory
Kinaesthetic	Olfactory

[1] *J. Mental Science*, vol. 100, 1955, p. 450.

The last synaesthesia, kinaesthetic–olfactory, was represented by a student who felt a real sense of swaying, as if on a ship, when smelling a sample of codliver oil. Other combinations have been reported since this paper was published, and of course the use of drugs such as mescalin and lysergic diethylamide (LSD) causes a much wider range of synaesthetic effects, possibly because more people become conscious of such experiences when under the influence of these psychotropic drugs.

Karwoski and Odbert[1] note that 'a preliminary survey of 274 college students revealed 165, or 60 per cent, who showed some tendency to associate colours with musical selections.' O. Ortman[2] quotes a figure of 28 per cent for subjects experiencing true visual-auditory synaesthesia. The differences in these figures need cause no surprise, as there is obviously a continuous gradation from real synaesthesia, where the induced impression is as intense and vivid as the stimulus, and at the other extreme, passing impressions only suggesting the ghost of another sense. The statistical analysis of these peculiarities of perception is always difficult, also, because the sense impressions have so few precise words to describe them. Visual effects have a fairly large vocabulary—the names of colours, shapes, geometrical figures and so on—but taste and smell can only be described in terms of food, flowers, or other 'concrete' objects with which we are familiar. For the perfumer or the flavour chemist, perhaps, many smells or tastes can be described in terms of animal, floral, fruit, spice, synthetic ingredients, and so on. But how does one describe a *new* smell or taste? In the case of the sense of touch, the difficulties are even greater and the vocabulary smaller: 'smooth' and 'rough' may denote a score of distinguishable. but indescribable, feelings, different again for the the child, manual worker, the mechanic, the hairdresser. . . .

In addition, there exists a prejudice that 'normal' or sane people should not be subject to such quirks of the brain. Galton

---

[1] 'A Review of Colour Music', *Psychological Reviews*, Ohio State University, 1946.
[2] *The Effects of Music*, loc. cit.

describes this censorship: '. . . the visionary tendency is much more common among sane people than is generally suspected. In early life, it seems to be a hard lesson to an imaginative child to distinguish between the real and the imaginary world. If the fantasies are habitually laughed at and otherwise discouraged, the child soon acquires the power of distinguishing them; any incongruity is quickly noted, the visions are found out and discredited, and are no further attended to. In this way the natural tendency to see them is blunted by repression. Therefore, when public opinion is of a matter-of-fact kind, the seers of visions keep quiet. . . .'

It is probable that the psychotropic drugs such as LSD act at least in part by inhibiting the action of the censors in the mind that prevent us from paying attention to our visions. Instead of ignoring chance resemblances between sense impressions, or repressing them from our conscious minds, such drugs induce people to take a renewed interest in such phenomena. Certainly synaesthesia is one of the commonly reported effects of the drugs, among other hallucinations.

The synaesthetic association between music and the other sense modes has long been recognized and discussed. Timbre especially has close colour affinities—the use of the words 'tone-colour' and 'Klangfarbe' shows an appreciation of the close linkage. Writers on orchestration such as Berlioz, Widor, Henry Wood and Aaron Copland use colour analogies freely in describing timbres or combinations of tone, and this feeling of a close resemblance between the two qualities is obviously very widespread. Percy Scholes, in the article 'Colour and Music' in the *Oxford Companion to Music*, writes:

'At a meeting of the Musical Association in 1876, Mr. R. H. M. Bosanquet, a distinguished scientist and researcher on the scientific side of music, suggested that, in order to make the reading of orchestral scores easier, different colours should be used in printing the staves devoted to various families of instruments. "Upon the point of colours he had found a remarkable agreement amongst musicians, and he suggested black for

strings and voices, red for brass and drums, and blue for wood."
Probably most musicians today would agree as to the propriety
of these associations, whereas very few would agree as to there
being any associative suitability in printing (say) the strings in
red and the brass in black or blue.'

Aaron Copland[1] shows his strong sense of colour association
in writing about orchestral characteristics of various composers.
For example: 'In place of brilliance, the neoclassic works
emphasized the dry sonorities of wind ensembles without the
string tone added—the greys and browns of a new and sober
colour scheme. Later, in the ballets of *Apollo* and *Orpheus*, Stravinsky evinced renewed interest in the strings, and gave them a
texture all his own; especially the string tone of *Orpheus* glows
with a rich, dark hue. . . .' Copland is not given to rhetorical
flourishes or a self-consciously decorated style in either his music
or his writing, so I conclude that these literally colourful
phrases come quite naturally to him, and parallel his own subjective feelings about the music he is describing.

Ortman in *The Effects of Music* quotes the example of one
subject with a very clear and precise association between
timbre and colour: the cello, for this individual, was indigo-blue,
the human voice green, the trumpet red, the clarinet yellow,
the oboe rose-red, the horn purple, the bassoon violet. Similarly
the painter Kandinsky saw flute tone as light blue, the cello
dark blue, the violin green, the trumpet red, drums vermilion,
church bells orange, and the horn and bassoon violet. There is,
for an agreeable change in this subject, a great deal of common
ground between the two subjects, though they were a generation and an ocean apart, and also between them and the colour
associations suggested by Mr. Bosanquet for scores. Certainly,
although I myself feel no kind of conscious colour associations
with music of any kind, the lists just quoted sound 'reasonable'
in some unexplained way.

Associations of colour and key have also excited great interest; Beethoven is said to have referred to the key of B minor

[1] *Music and Imagination*, New American Library, 1959.

as 'black', and many musicians have similar associations. (If 'black' is meant to be a symbol of death and grief and the other dark emotions, Beethoven's use of the key of B minor is hardly as dark and tragic as in J. S. Bach, to whom this key was obviously of great significance. Some of Bach's most despairing and moving music is in B minor, and it is reasonable to imagine, to quote a modern use of the key, that the lonely, wistful fugue in B in Hindemith's *Ludus Tonalis* was probably suggested by, and a tribute to, the great B minor fugue at the end of the first book of the *48*, where the expression is so intense that one almost feels the breaking-up of Bach's chromatic harmony into a tonal chaos.) Rimsky-Korsakov and Scriabin, as befits such brilliant orchestral colourists, apparently had strong associations between key and hue:

Key	Rimsky-Korsakov	Scriabin
C major	White	Red
G major	Brownish-gold	Orange-rose
D major	Yellow, sunny	Yellow, brilliant
A major	Rosy, clear	Green
E major	Blue, sapphire, sparkling	Bluish-white
B major	Sombre, dark blue shot with steel	Bluish-white
F♯ major	Greyish-green	Bright blue
D♭ major	Dusky, warm	Violet
A♭ major	Greyish-violet	Purple-violet
E♭ major	Dark, gloomy, bluish-grey	Steel colour with a metallic lustre
B♭ major	—	Steel colour with a metallic lustre
F major	Green	Red

The disagreements, as with the other synaesthetic effects, are unimportant. We are not concerned with the question of everyone having the same associations, but with the fact that such associations exist and are very real to the people who experience them. In fact, they are so real that they tend to generate a curious intolerance. While a person who sees, say, A major as rosy, can just about comprehend that many other people manage to exist without such associated colours, he tends to be far more impatient with the person who sees A major as green. In the one case, it is obviously merely a loss to the people who are unable to see the appropriate colours, but to see a rosy key as green must seem to verge on wilful wrong-headedness. Galton draws attention to this curious reaction among people who associate different colours with vowel sounds, or see a different 'number-form', and so on. (A number-form is a visual association where the numbers appear to have a specific layout in space: this is used by many people as a positive aid to arithmetical calculations, or for remembering dates, etc. These again can be so real that Galton quotes the example of Mr. Flinders Petrie who could imagine a slide-rule so vividly that he could carry out calculations on it. Some of the calculating prodigies appear to have possessed similar aids to their manipulation of numbers.)

Visual-auditory associations of musical pitch with various colours have, curiously enough, not been recorded very frequently. I say curiously enough, because there is a large amount of writing on this subject because of a supposed analogy between the seven colours of the spectrum, with their steadily increasing electromagnetic frequency from red at one end of the spectrum to blue or violet at the other, and the seven notes of the diatonic scale, with their increasing audio-frequency as the pitch rises. The analogy between the two 'scales' has been of interest ever since Newton made clear the nature of the colours of the rainbow: Louis Bertrand Castel invented an Ocular Harpsichord as early as 1739. In this the notes of the scale, C to B, produced colour harmonies with coloured ribbons. Newton himself pointed out the resemblance between the colour scale and the musical scale when he announced his discovery of the dispersing power of the prism.

For readers who are interested, the detailed analogy between the musical and colour scales can be calculated as follows: taking a typical base-note such as c', the ratios of the frequencies of d', e', and so on may be taken in the Pythagorean ratios of 9:8, 5:4, etc. If c' is taken as equivalent to a typical frequency for red light, say $4\cdot40 \times 10^{14}$ c.p.s., then d' must be equivalent to $4\cdot40 \times 10^{14} \times \frac{9}{8}$ c.p.s., which is in the orange range, and so on. The complete scale works out as follows:

	Ratio to c'	Corresponding Light Frequency	Approximate Colour
c'	$1:1$	$4\cdot40 \times 10^{14}$	Red
a'	$9:8$	$4\cdot95 \times 10^{14}$	Orange
e'	$5:4$	$5\cdot50 \times 10^{14}$	Yellow
f'	$4:3$	$5\cdot86 \times 10^{14}$	Greenish-yellow
g'	$3:2$	$6\cdot60 \times 10^{14}$	Blue
a'	$5:3$	$7\cdot33 \times 10^{14}$	Indigo-violet
b'	$15:8$	$8\cdot25 \times 10^{14}$	Ultra-violet (invisible)

Having said as much, little more can be said. The analogy breaks down because there is absolutely no sense in which the 'octave' of red, which would be light with a frequency of $8\cdot80 \times 10^{14}$ c.p.s., has any similarity to red. c'' is so close in character to c' that we think of them as being the 'same note' in many ways: the light at $8\cdot80 \times 10^{14}$ cycles is an invisible ray, liable to blister our skins, and quite unlike any kind of visible light. The same applies to the 'octave below': $2\cdot20 \times 10^{14}$ c.p.s. is also an invisible ray of light in the infra-red region.

The other difficulty about this colour/sound analogy is that harmony in colour terms is not a precise matter as it is in musical terms. Musically, c', e', g' is a concord, while c', $f\sharp'$, g' is a violent discord. In colour equivalents these correspond to red-yellow-blue and red-green-blue respectively, and whatever one's sensitivity to colour it is unconvincing to say that one of these combinations is concordant and the other discordant.

Turning from visual-auditory synaesthesia, it is interesting to consider other senses which can be influenced by sound. McKellar quotes a typical tactile-auditory description from one of his subjects: 'A trumpet sound is the feel of some sort of plastics, like touching certain sorts of plastic cloth—smooth and shiny—I feel it slipping.' This sort of touch association is particularly appropriate to some tone-colours. I have heard the flute fluttertongue passage at the beginning of Bartók's *Concerto for Orchestra* described as 'furry', and the comment seems intelligible.

A reversal of this type of reaction, in which a tactile or body-schema impression evokes a musical image, is also quoted by McKellar. A musician, during a severe illness and after several uncomfortable nights, said there was 'only one position in the bed, which is in F♯ major, and that is the only one which makes sleep possible!' As a clue to this highly idiosyncratic method of description, it occurred to me spontaneously, when I first read this passage, that the musician must have meant that he was lying diagonally across the bed. When I came to analyse this quite unconscious interpretation, I decided that it was partly conditioned by the lozenge shape of the sharp symbol itself, but mainly by the way that the key signature of F♯ major straddles the stave in a diagonal direction.

Despite the difficulties of description, body schema, tactile and kinaesthetic synaesthesias and hallucinations seem fairly common, especially in the form of the hypnagogic 'falling' experience, coming to the mind on the verge of sleep. Synaesthesias in the strict sense of the word, where the induced image is as vivid and intense as the 'real' stimulus, are not rare experiences. McKellar's group of 182 students contained 21 per cent of individuals with experience of synaesthesia. The specific visual-auditory 'colour music' group of sensations has been estimated variously, as has been said, as 60 per cent, 28 per cent or as low as 16 per cent, according to the definition. M. D. Vernon[1], who quoted the last figure, obviously had very precise ideas of the difference between synaesthesia proper and more minor associations. Writing of her investigations by questionnaire she says, 'Of those who answered . . . 16 per cent had some more or less well-established connection between colour and music, 4 per cent had temperature and 2 per cent taste or smell associations, 10 per cent characterized keys, 8 per cent more

[1] 'Synaesthesia in Music', *Psyche*, vol. 10, 1929–30.

had a sense of the individuality of keys while listening, 12 per cent had various kinaesthetic patterns and 18 per cent visualized actual notes. Naturally many of these groups overlapped, most of my subjects combined two or more of these characteristics. Excluding visualizers of notes, the total was 30 per cent, including them, 40 per cent. I can find no statistically significant differences as regards musicalness or unmusicalness.'

Two of the subjects in Vernon's investigation, who normally experienced synaesthesia, were completely unable to understand how other people who did not have this sense could perceive any pleasure in music at all. This is reminiscent of the attitudes of some of Galton's correspondents. Vivid synaesthesia can sometimes lead to unpleasant experiences, as in the case of the remarkable memorizer S. V. Shereshevskii, who had powers of association which enabled him to memorize almost limitless amounts of information, even nonsense words or meaningless mathematical formulae. He found that, at lunch in a restaurant, the taste of the food varied with the music which the orchestra was playing ('probably they're playing specially to improve the taste'), but noises from the roof of the restaurant when repairs were going on made the lunch suddenly taste 'horrible, completely spoiled'.[1]

Less vivid impressions, the passing suggestion of images, parallel sense impressions, and ripples in one sense mode caused by a disturbance in another, must be very widespread. The whole language of metaphor and symbol is shimmering with examples, ranging from phrases in daily use to the most fantastic and personal conceits. Very often the everyday coinages have become so hackneyed that we hardly notice that they are metaphors at all: a 'dry' or 'smooth' wine, a 'loud' suit, a 'flat' taste and so on. At the other end of the scale, poets tend to slide into their own private synaesthetic images without realizing, or possibly caring, that these are not universally understood. Edith Sitwell's 'Emily-coloured primulas' and 'the morning light creaks down again' are typical of this private language: Hoff-

[1] I. M. L. Hunter, *Memory*, Penguin Books, London, 1964.

74

mann's Johannes Kreisler, 'the little man in a coat the colour of C♯ minor with an E major coloured collar' suggests very little to most people unless they have key-colour associations. Such intense dependence on a private language may seem rather arrogant in a professional writer, but it may well be that they assume, as did many of Galton's subjects, that such images are universal. Had Mrs. Haweis, one of his correspondents, been a fashionable writer, we might have seen some really remarkable productions in the way of extravagant metaphors:

'Printed words have always had faces for me; they had definite expressions, and certain faces made me think of certain words. The words had *no* connection with these except sometimes by accident. The instances I give are few and ridiculous. When I think of the word Beast, it has a face something like a gurgoyle. The word Green has also a gurgoyle face with the addition of big teeth. The word Blue blinks and looks silly, and turns to the right. . . .'

In general, apart from such delightfully personal instances as this, synaesthetic description adds greatly to the richness of metaphor. Edith Sitwell's lines from *Façade*:

> *The fire is furry like a bear,*
> *And the flames purr . . .*

owes its charm not only to the alliteration but also to its tactile imagery. Keats, as Robert Graves has pointed out, also made great use of tactile images, but he was also strongly attracted to synaesthetic images connected with taste and smell.

> *Upon the honeyed middle of the night . . .*
> > *The Eve of St. Agnes*
> *Yet I must dote upon thee—call thee sweet,*
> *Sweeter by far than Hybla's honied roses . . .*
> > *Sonnet to* xxxxxx

(We forget sometimes that the use of the word *sweet* for a person is a synaesthetic description. It is one which has become commonplace.)

75

Now I have tasted her sweet soul to the core . . .
Endymion, II
Bitter coolness . . . *Endymion*, IV
Tasting of Flora and the country green . . .
Ode to a Nightingale
. . . *There she lay,*
Sweet as a muskrose upon new-mown hay . . .
Endymion, IV

Conclusion
Enough has been said to indicate that synaesthesia is a reasonably common experience connecting all the sense modes, and that these fused or parallel associations add a great deal to the richness of our mental life; such associations are obviously of great importance in the development and our interpretation of metaphor and symbol.

We saw in the last chapter how the brain processes all its incoming information, and such 'internal' store material as memory, in terms of a common code of electrical pulses which are used for all its operations, and which are exactly the same for all nervous events. Normally we know that a coded signal is 'sound' because it comes from the nerves connected to the ear, and that a signal, using the same pulses, is 'sight' because it comes from the eye.

If we think now of the process of *identifying* the signal, we are faced with a more complex system. Suppose our ears are agitated by the sound of a bassoon. The complex of electrical pulses is sent to the auditory cortex, but at this point we are only aware that something is happening in our ears: the signal has not been interpreted. To find out what exactly is happening, we have to compare the signal with a store of signals kept somewhere in the brain, and eventually this processing stage reports that there is identity of pattern in many respects between this particular pattern of pulses and previous ones which we have learned to call 'bassoon tone-colour'. This seems a very slow process, as described here, but it is, of course, done in a fraction of a second, as with all these comparison processes, of which the brain carries out millions every hour.

The pattern of harmonics and so on which makes up 'bassoon tone-colour' has of necessity been translated into electrical pulses. Now the data-processing stages of the brain have no direct knowledge of the nature of bassoon tone: they merely identify the pattern of pulses as being similar to other sets of patterns which we have previously labelled 'bassoon', out of the billions of patterns kept in the memory store of the brain. Just as skilled film editors can tell by looking at the shading of an optical sound track exactly when the strings come in or a trumpet call stops—without ever 'hearing' the music—so the brain must work in a similar way in its comparison operations.

However, like most universal codes, that of the brain can suffer from ambiguity. Two different messages may turn up in the brain as patterns of pulses so similar that we confuse the two meanings. This happens frequently with words, of course. We tend to forget that written words are a code for conveying our meanings until we run up against a problem sentence like 'Mrs. X wishes to announce that she has cast off clothing of every description,' and similar ambiguities. While, in fact, the processing part of the brain may be comparing a large group of past memories with the new signal which has just come in, it may find resemblances with several patterns in the 'store'. If our bassoon tone is compared with past memories of bassoons and found similar, we think to ourselves, almost unconsciously, 'that is a bassoon', and think no more of such an everyday occurrence. If, however, our brains also perceive and report a resemblance between the incoming pattern and a previously remembered *colour* pattern, as Kandinsky's brain must have done, we have a sudden insight of the equivalence between bassoon tone and the colour violet, or of some other synaesthesia.

While synaesthesia of this type is not universal, we can see it as the extreme case, vivid enough to be reported as something unusual, of a great sympathy between the various sense modes which we take for granted in the form of metaphor and symbol.

77

When the incoming signal, as in the case of music, is a pattern which has no immediate function as a *useful* sense impression, in the sense that we do not, as with many visual impressions, or more mundane auditory impressions like the noise of an approaching car, have to do anything as a response, it seems more likely that the brain will turn up a number of equivalent patterns from other sense modes. While the notes themselves may have nothing to say to us, the electrical patterns into which they are translated in the brain may be compared and identified with other patterns from other sources.

Frank Howes[1] has written a passage which states, with remarkable clarity, a musician's arrival at the same theory which I had reached from other considerations:

'Any competent musician can design patterns of tone, but it is only the true composer who can say something with them, and he does so by an alchemy of the mind, the fundamental nature of which would seem to be a kind of analogy. By virtue of some recondite likeness between two kinds of experience in two different sensory media, a relationship is established. . . .'

This is, in a few words, the thesis I have tried to develop to account for the actual expressive nature of music. From modern neurophysiological theory I have tried to throw some light on the nature of the 'recondite likeness' which may exist between two or more sense modes, and to show that such likenesses may occur outside the world of art, and are, indeed, a common feature of brain activity. Much remains to be done before any analysis of the patterns actually passing through the brain can be carried out, and for this reason my thesis must remain a hypothesis in many respects. However, in the next chapter I have attempted to analyse some of the patterns which could be the subjects of such analogical working in the brain, and to pinpoint those which may be of the greatest interest for any future investigation.

[1] *Man, Mind and Music*, Secker and Warburg, London, 1948.

5 The Patterns

In Chapter Two, the basic patterns of musical experience have been described, and it is clear that they can be analysed into patterns of tensions and resolutions. These patterns are obviously more subtle and varied than such a shorthand description can convey, and they cover melodic, harmonic, rhythmic and dynamic tensions and every combination of the four elements, but, just as the harmonic analysis of a complex wave-form into its component sine-waves may give us useful information about the constitution of the complex, so the analysis of music into its component patterns can tell us a great deal about the nature of music, without detracting in any way from our pleasure in the whole. The units of the patterns may be as short and simple as the two-chord tension/resolution of a cadence, or they may add up to a structure as elaborate and highly-developed as the twenty-one minutes of Sibelius's Seventh Symphony, but the general pattern of tension followed by resolution remains the same in essence.

The various methods of forming, shaping, and balancing these patterns constitute the art of composition. This has obviously grown up largely by trial and error. A composer 'feels' what is the right pattern for his own personal communication by rejecting the patterns which seem to him, introspectively, to be wrong, and holding on to the ones that are right. Also, with the help of theorists, teachers and critics, he recognizes and experiments with those elements which seem 'right' in the compositions of other people. The degree to which this can be taught varies with the composer and the teacher. A composer with a strong will to communicate will learn the best methods by his own efforts and his own studies of other men, and he will sort out the successful procedures for himself. Walton and Elgar obviously worked this way, and Mozart and Mendelssohn were composing competently before they undertook the normal course of theoretical analysis as an aid to composition. It is not my purpose to dwell on the teaching of music, merely to clarify the point that strong, successful, meaningful patterns for music cannot be constructed on any basis save that they give pleasure to the composer first, and afterwards to his audience. It is also worth saying at this point that this thesis aims to analyse the music which has 'worked' in the past, and to see how it works, rather than to suggest any new procedure for the future.

In the brain, the patterns of tension and resolution implicit in music are translated into an electrical code of pulses, used in common for all the senses, including hearing and bodily sensations. This *lingua franca* of the brain, as we have seen, can give rise to associations, or double or multiple sensations in different sense-modes. In the last chapter some of these multiple associations were considered in the extreme form of synaesthesia, and also in the more common forms of a sympathy between various sense-modes heading to a general identification of similar patterns in any of the several human senses.

From the incoming musical pattern, therefore, the brain makes an attempt to identify the consequent electrical pulse pattern. If the pattern of tensions and resolutions in the music is similar to that usually connected with body movement, touch, sight, taste, or any other sense impressions, we will have a sense of association between the two or more sets of impressions amounting, to a greater or lesser degree, to identification. In fact, if we consider the tensions and resolutions communicated to us in music, it is clear that they are only examples or representatives of a far larger class of similar experiences—all circulating in our brains in the same electrical language, and all reduced to the same terms. Hunger and thirst followed by satisfaction, pain and its relief, expectation culminating in the arrival of the desired object, sexual excitement and its fulfilment, bowel retention and evacuation, all have their own patterns, yet with a family likeness which is unmistakable.

Some of these impressions form our earliest experiences: before our eyes have learned to focus or our fingers to grasp, we have experienced one particular cycle over and over again—tension from lack of oxygen in the blood, followed by relief as we breathe in, then tension from the effort of raising the rib-cage, followed by relaxation as we lower it to breathe out. Consciously or not, we shall go on experiencing this particular pattern over six million times every year of our lives. Breathing has, in many parts of the world, retained the sense of mystery, of being the root of life, which we tend to forget in the West. *Ātman* in Sanskrit means both breath and spirit (and the word has passed down into modern German with very little change after three millenia). One of the most important yoga exercises for obtaining control, not only of the body, but of consciousness, is *prānāyāma*, a method of controlling the breathing so as to make it entirely rhythmical, and at the same time deep and physiologically efficient. Bhoja (*Yoga-sūtras*, I, 84) remarks: 'All the functions of the organs being preceded by that of respiration—there being always a connection between respiration and consciousness in their respective functions—respiration, when all the functions of the organs are suspended, realizes concentration of consciousness in a single object.'

Eliade[1] puts this point at greater length: 'Through *prānāyāma* the yogin seeks to attain direct knowledge of the pulsation of his own life, the organic energy discharged by inhalation and exhalation. *Prānāyāma*, we should say, is an attention directed upon one's organic life, a knowledge through action, a calm and lucid entrance into the very essence of life. Yoga counsels its disciples to live, but not to abandon themselves to life. The activities of the senses possess man, corrupt and disintegrate him. In the first days of practice, concentration on the vital function of respiration produces an inexpressible sensation of harmony, a rhythmic and melodic plenitude, a levelling of all physiological unevennesses. Later it brings an absolute feeling of presence in one's own

---

[1] *Yoga—Immortality and Freedom*, trans. Trask, Pantheon Books, 1958.

body, a calm consciousness of one's own greatness. Obviously, these are simple data, accessible to everyone, experienced by all who attempt this preliminary discipline of respiration. Professor Stcherbatsky writes that, according to O. Rosenberg, who tried certain yogic exercises in a Japanese monastery, this agreeable sensation could be compared "to music, especially when one plays it oneself." '

It is interesting to note that the yogic control of respiration, in *prāṇāyāma*, produces sensations which are best described in musical metaphors such as harmony, rhythmic and melodic plenitude, and similar phrases which are in fact common in the literature of yoga. This is evidence of the way in which the sensations of a bodily function and the patterns of music are felt instinctively to be equivalent. The mystical approach to breathing which is a feature of yogic writing may seem overdone at first, but it is a useful reminder of the fundamental part which breathing plays in our lives, a part often forgotten because it is so habitual.

This mystical approach to breathing and breath control as a means to mental harmony and enlightenment is not confined to classical yogic writing. Similar respiratory exercises are practised in Taoism (*t'ai-chi* = 'embryonic respiration') and Islamic mystical practice (*dhikr*, which is strikingly similar in its methods and reported results to *prāṇāyāma*). Nicephorus, in the thirteenth century, wrote of similar exercises practised by Christian mystics, and expressed the view that these could 'unite the mind with the soul'. An even more complex example of the effect of such exercises in lowering the barriers between various sense-modes is given by the modern writer Muhammad Anūn al-Kurdi, who describes the exercise of the '*dhikr* of the tongue', in which the *dhakir* holds his tongue tightly against the soft palate and holds his breath; among adepts, this exercise brings about auditory hallucinations with great regularity.

The most obvious example of the connection between respiration and music is singing. There are some melodies which fit the rhythms of breathing so exactly that it is difficult to imagine them performed on anything else but the human voice, and in most cases it will be found that they contrive to make their points of emotional tension coincide with points of physical tension for the singer. The phrases fall naturally into the limits of one breath and tense phrases 'stretch' the capabilities of the singer by making him use his lungful of air to its utmost extent. The great tradition of Italian opera has always exploited this technique to the full, making breath control one of the aesthetic patterns at the composer's disposal to express his emotional content. If an inept singer breaks one of these well-constructed phrases to take a breath which was not considered necessary by the composer, we resent this, not so much because of the mechanical failure—the art of singing is not a branch of athletics—but because he has destroyed one of the patterns which go to make up the total musical synthesis.

Wind instruments, and in particular the horn, lend themselves to this type of breath phrasing, and the horn part of Britten's *Serenade*, for example, is rich in breath rhythms, with a corresponding physical effect on the listener which must have contributed greatly to the popularity of this work. Composers who are wind players themselves, such as Holst and Hindemith, or singers, often show a great feeling for this aspect of musical pattern-construction, but it is obviously not confined to the practitioners of the instruments: in many ways we may assume that experience of breathing and breath patterns is enough to enable a sensitive composer both to write convincing patterns of this kind in his music and to write effective wind parts which take account of the breathing problems of the players.

A striking piece of breath rhythm occurs in the Overture and the second act 'stone guest' scene in *Don Giovanni*: the scale

passages which add to the terror of the confrontation between Giovanni and the statue of the Commendatore fall exactly into the rhythm: inspiration (rising scale), exhalation (falling scale).

In a typical performance these patterns occurred at the rate of approximately 14 per minute, which is characteristic of the rhythm of rather quickened breathing, as in stress or fear. The pace is maintained throughout the statue scene, through the iambic section, until the opening of the pit at Giovanni's feet causes an increase in speed beyond the range of breath rhythms. Similar instrumental examples may be found in many works; the opening of Brahms's Fourth Symphony falls naturally into breath rhythms, and in *Daphnis and Chlöe* Ravel has written a passage that actually sounds rather like a sigh itself, and certainly follows the pattern of one exactly.

Apart from the rhythm of breathing, the other dominant rhythmic sound in our lives is the heartbeat. Perhaps we become less aware of it in adult life, unless we run after trains or go rock-climbing, but its influence extends back long before birth, to a time when we had more leisure to listen to the heartbeat, and fewer distractions. Our mother's heartbeat is the first sound which we hear, and as babies at the breast the same rhythm also: we hear it when other sense-impressions are fragmentary or non-existent, so that it assumes a far larger place in our consciousness than most other sense-impressions experienced afterwards, which have to compete for attention with an enormous number of other pieces of incoming information. It is significant that the (admittedly limited) work carried out on the subject suggests that mothers tend to rest their babies' heads rather more frequently on the left side of the breast than the right, and that this does not depend on the left- or right-handedness of the mother. Perhaps mothers have learned from experience that the baby is more easily soothed if it is placed near the sound of the familiar heartbeat.

Apart from this aspect of the heartbeat, how does it fall into the tension/resolution pattern? Lorus and Margery Milne, in

84

their *The Senses of Animals and Men*[1] give a clue to the possible origin of the tensions in an account of some experiments by Dr Lee Salk.

'The soft lub-dupp, lub-dupp of a relaxed mother's heartbeat was played recently over the loudspeaker system into a nursery room full of new-born babies. . . . Most of them soon went off to sleep. The rest appeared reasonably contented. The recording stopped. Within a few seconds a good many babies woke up: some began to cry. Then a new record was played: the rapid heartbeat from an excited woman. The sound was no louder, but all of the sleeping babies awoke immediately. Every infant grew tense, as though in fear. When the first recording was played again, peace spread through the nursery.'

In fact, tempi which are faster than the normal heartbeat, or more irregular, act as tensions in themselves. The irregularities and their tensioning effect probably account for the stimulating effect of dotted rhythms compared to the placid and soothing effect of the same passages in equal notes. Where the dotted rhythm is combined with tense harmony and tempo faster than normal breathing, as in the *Don Giovanni* example we have already considered, the effect is not only stimulating but definitely frightening.

The last dominant influence on our rhythmic sense, at least in our interpretation of small units of pattern, has already been mentioned. This is the 'minimum reaction period' of about 0·7 second, which is the time necessary for the brain to receive a message, identify the content by comparing it with previously remembered experience, which we have called the 'data-processing' stage, and react to the message in some appropriate way. This neurophysiological metronome sets a *tempo giusto* at about M.M. 80–6 which acts as a norm: slower or faster tempi than this appear as tensions.

The rhythms of the heart, lungs and brain account for most of the small units of pattern to which we have grown accustomed. Larger units with more subtle patterns are accounted for by the

[1] André Deutsch, London, 1963.

characteristic tension/resolution patterns of pain, followed by its relief, fear, followed by running away or fight, bowel action, sexual excitement leading to the tremendous tensions of the climax, followed by relaxation: among the mental patterns are expectation followed by fulfilment or, alternatively, by the acceptance of disappointment, grief, followed by consolation, and so on. Many of these have two possible types of pattern. There solution may come as an explosion of action, or as a slow relief of tension, as Koestler has pointed out in *The Act of Creation*.[1]

The explosion may be laughter at a comic situation (which has previously created tension by its juxtaposition in the mind of two apparently alien, irreconcilable factors), sudden action after the tension of fear (fight or flight), sexual activity after desire. The slow release may be quiet satisfaction at solving some previously puzzling problem or situation, or the slow evaporation of fear when we realize that there was really nothing to be frightened of, or the cooling of desire when we realize that its object is unapproachable.

In any case, the larger patterns are more variable and complex, but they still have their characteristic tension/resolution relationships, and these have at least a family resemblance. Indeed, it seems likely that the occasions when a pattern comes out almost exactly the same as a previous and remembered one account for the feelings of 'I have been here before'. The more basic the emotion or bodily function, the more often the pattern is repeated, and this applies particularly to those events which happen in childhood.

It is not necessary to go as far as the orthodox Freudian psycho-analytical school, in ascribing our adult behaviour entirely to the results of childhood experiences, to see how deep an impression is stamped in us by the patterns of hunger, fear, pain, bowel retention, discomfort and similar tensions: at the earliest period of our lives they *are* our lives, providing the material for the whole of our experience and our thought, and they continue

[1] Hutchinson, London, 1964.

to occur, day after day, until we reach the final resolution of all our tensions.

This, then, gives us the first step in explaining the 'meaning' of musical patterns. While they are strictly speaking meaningless in themselves, the fact that they are translated in the brain into the general *lingua franca* of all other patterns—mental patterns such as grief, expectation, fear, desire, and so forth, and bodily patterns such as hunger, pain, retention, sexual excitement, any of the tensions associated with a raising of the adrenalin level in the blood—and the corresponding resolutions—allows us to see the similarities between the musical patterns and these more personal ones which form the constant undercurrent of our thoughts.

The momentary tension of a discord, so alien to our experience if we regard it merely as a collection of sound waves, in the brain becomes one of an enormous family of similar tensions, and we are aware of the analogy between the discord and other minor tensions: such as a catch of the breath, the effort of a muscle, a slight mental check in solving a problem. A long-held minor 6–5 appoggiatura becomes akin to a long-held breath, followed by its release, or the effort to move a heavy weight, followed by the relaxation of the muscles after it has been moved.

Suzanne Langer[1] makes the same point: 'The tonal structures we call music bear a close logical similarity to the forms of human feeling—forms of growth and attenuation, flowing and slowing, conflict and resolution, speed and arrest, excitement and calm, or subtle activation and dreamy lapses—not joy or sorrow perhaps, but the poignancy of either or both—the greatness and brevity and eternal passing of everything vitally felt. Such is the pattern, or logical form, of sentience; the pattern of music is that same form worked out in pure measured sound and silence. Music is a tonal analogue of emotive life.' Frank Howes has extended this theme of equivalence[2] in an excellent series of lectures owing much (on this point) to Langer's writings.

[1] Suzanne Langer, *Form and Feeling*, Routledge, London, 1953.
[2] Frank Howes, *Music and its Meaning*, University Press, London, 1958.

It is not necessary for the analogy to be perfect in every detail. The brain has a great capacity for sorting out the essential points from a mass of detail, so that, for example, we can recognize the pattern 'horse' whether the animal is grey or chestnut or piebald, whether we see it very small on the horizon or very large near at hand, whether we see it foreshortened as it leaps over a fence towards us, or in silhouette, formalized, on a crest or coat of arms. In every case the shape, colour and size projected on to the retina of our eye may be different, but we are still able, it seems instantaneously, to recognize the 'horse pattern' which underlies all these varied views. This accomplishment of recognizing objects despite the fact that they can present different visual images (as in the simple case of a plate, which is circular if we look at it direct, but elliptical if we look at it at an angle) is something which has to be acquired, but we seem to acquire it very early in life, usually during the first twelve months.[1]

Among these perceptual acquisitions, we learn to recognize the pattern of a tune whether it is played slowly or quickly, by a large orchestra or on a penny whistle, at any pitch and in any key, even, as in some contrapuntal devices, when it is played upside down. The sounds which come from the $E\flat$ clarinet in the 'Witches' Sabbath' movement of Berlioz's *Fantastic Symphony*, if recorded and drawn out by a wave-analyser, would seem to have very little relation at all to the opening theme similarly analysed, yet the ear detects the *idée fixe* immediately. Think of the pattern 'bassoon tone', which we considered in the last chapter: if we decide to listen for the bassoon in an orchestral work, our ears will appear to develop a sudden sensitivity to the doings of that particular instrument, even though it may be playing only some subordinate and quiet accompaniment figure in the midst of a mass of other sound. In fact, the brain's powers of selecting those patterns which it wants to examine may even extend to control over the sense organs themselves, at least to the extent that the electrical signals from those organs

[1] M. D. Vernon, *The Psychology of Perception*, Penguin Books, London, 1962.

to the brain can be modified. Galambos[1] carried out experiments with cats, on this aspect of attention. The cats' auditory nerves were tapped so that the electrical pulses passing from the cochlea to the brain were also picked up by an amplifier, and recorded. Obviously any sound would cause an easily identified signal to pass along the nerve, and a steady series of clicks from a metronome caused a regular series of signals. With these conditions made familiar to the cat, it was not long before the signals were completely regular and predictable—a steady state had been set up. Now the cat was shown a mouse (protected by a glass jar): immediately the activity in the auditory nerve diminished or ceased altogether, despite the fact that the metronome was still clicking away. The cats turned a literally deaf ear to the metronome when anything as important as a mouse was presented for their attention. Similar experiments were carried out by Hernandez-Peon, Scherrer and Jouvet,[2] who placed electrodes in the brainstem at the point where the auditory nerves connect to the brain: they also found that a mouse was more important to a cat than a metronome, and even the *smell* of fish caused the 'deaf ear'.

The ability to extract the important pattern from a wealth of extraneous detail, to see the essentials of a situation stripped of superficial differences, and to generalize from experience, is the key to the whole coherency of our mental life. Without this ability our lives would be spent in an incomprehensible, noisy, dazzling chaos—the 'big, booming, buzzing confusion' which William James described as the world of the new-born infant, who has not had the opportunity to acquire any ideas of generalizing from experience. We tend to forget how much we have had to learn since that early stage: to take a simple example, my hand is lying in front of me presenting a view of four fingers and a thumb clearly visible and separate. Let us analyse how I know it is my hand. I know it is mine because on countless former occasions I have experienced the sense of touch coming

[1] *J. Neurophysiol.*, vol. 19, 1956, p. 424.
[2] *Science*, vol. 123, 1956, p. 331.

from it, because it is fastened to my arm, because it is always there when I am. I know it is *a hand* because my parents taught me that the odd starfish-shaped object on the end of my arm was called 'hand', and I have found from experience that other people (in this country) also call similar things 'hands'.

Now, if I turn it sideways, I can no longer see the four fingers as separate things, and the shape is less like that of a starfish, and more like that of a claw. From all visual points of view it is a different object—it is only because I have in my brain an enormous memory store and the ability to generalize from experience that such a fundamental thing as recognizing a part of my own body is possible.

The actual mechanism of inspection, comparison and extraction of 'similarities', which forms the basis of memory and generalization, is still a matter of discussion, but the general nature of the operations is clear. Pringle[1] describes memory traces as 'coupled resonators', an evocative phrase which describes well many of the operations of matching up experiences and concentrating special attention on their similarities. If we ask what sorts of things 'match', the answer must be 'it depends what you are looking for'. The brain has countless different frames of reference against which to compare a new experience. If I am looking for blue beads in a bag full of mixed ones, each bead on which my eye rests is assessed against the frame of reference 'colour', and all blue objects 'match' in this case, or are at equipotential. To use a metaphor which may have no physical counterpart in the brain, but is useful for describing the type of process involved, some beads pass the 'blue' filter. If, on the other hand, I am trying to sort out large beads from small ones, only the 'relative size' frame of reference comes into use, and colour is of no importance. Often very oddly assorted things have similarities in certain frames of reference: if I am looking for triangles, the percussion instrument, a picture of a bell tent, a jam turnover and the three-person situation so beloved of dramatists may all pass through the same filter—the triangle.

[1] *Behaviour*, vol. 3, 1951, p. 174.

Koestler bases his theory of the mechanism of discovery, the 'Eureka-act', on the idea that inspiration or discovery, both in the arts and the sciences, arises from seeing that there is a similarity in some frame of reference between two concepts which were previously thought to have nothing at all in common. They both pass the same filter, if we have the right filter ready. Archimedes, who gives a name to the process, was faced with the problem of finding the volume of an intricate crown, made for Hieron of Syracuse, so that he could assess the amounts of gold and silver used in it by the measurement of its density. Obviously he could not destroy the crown by melting it down and shaping it into a cube, or sphere, or other shape of calculable volume. The story goes that he suddenly realized, in his bath, that the rise in the water level of the bath as he got in was a measurement of the volume of his equally intricate body. A bath, a habit as familiar as the evening meal, had suddenly found its way through the filter 'measurement of volume', and appeared in an entirely new context: or perhaps we should say that Archimedes's body and the Tyrant's crown, hitherto quite separated in Archimedes's mind, had suddenly appeared at equipotential as complex solids which could be mensurated by immersion in water.

Many problems depend for their successful solution on the selection of the right filter, frame of reference or thought-matrix, as Koestler phrases it. The riddles of the Sphinx or Turandot depend on finding one object which can pass several filters in different contexts, like the curiously-cut pieces of cork, beloved of children's puzzle-book compilers, which can pass through a circular, square or triangular hole. Euler's famous problem of the Königsberg bridges, the beginning of topological mathematics, depends for its solution on seeing the town bereft of all the extraneous detail of houses and areas, and only considering the relationships of the bridges themselves.

With such a system of comparison, generalization and analysis going on in the brain it is not surprising that musical patterns can pass through the filters that we normally associate with the patterns of our bodily and mental life. There is a Eureka-act, as

Koestler would say, between the musical pattern and the human pattern, and we are suddenly struck by the similarity, the equipotential. 'The world becomes akin to us through this power to see in forms the joy and sorrow of existence that they hide; there is no shape so alien that our fancy cannot sympathetically enter into it. Unquestionably the vividness of these perceptions is increased by our abiding remembrance of the activities of our own bodies.'[1] The words are taken from an earlier philosopher, but here is the germ of truth which underlies, and makes palatable, a great many of the self-contradictory theories of Einfühlung, or Empathy, as expounded by Lipps and his followers. To say that aesthetic pleasure is an enjoyment of our own activity in an object—a fair statement of Lipps' position—leads to various difficulties, especially when our feelings 'in' a building or landscape are concerned. However, if the mechanism is that the shapes and patterns in the architecture or the landscape are translated into the brain's electrical code, and in this form show similarities to memory traces of bodily or mental activities, then the difficulties seem less, and our intuitive belief that these theories contain a basic truth is justified.

In the rather arid language of the professional psychologist, M. H. Ward[2] puts a similar point of view: 'Music, acoustically, is nothing more than sound frequency pattern, combined with rhythmic pattern. . . . That these rhythm-pitch patterns do have a profound effect on us, that they do indicate that the composer that chose to set them to paper did express a "gay" or "disturbed" or a "gracious" emotion at the time they were conceived, obviously must be due to some association between these patterns and the human mind. The answer is to be found in the fact that the organism also passes through a series of rhythm-pitch patterns as it travels through the emotional gamut of its existence, and that the passages played by one or more of a group of instruments bring these to mind.'

[1] Lotze, *Mikrokosmos*, vol. 5, ii, 1856–65.
[2] 'Psychomusic and Musical Group Therapy', *Sociometry*, New York, vol. 8, 1945, p. 238.

Vernon Lee commented in *Beauty and Ugliness, and other studies in psychological aesthetics*:[1] 'We attribute to lines not only balance, direction, velocity, but also thrust, strain, feeling, intention and character.' This is also true for musical 'lines'. Deryck Cooke, in *The Language of Music*, has analysed in detail how the elements of pitch can be transposed from 'up and down' to 'out and in', 'away and back', and more complex emotional movements. 'Thus, at the opening of the Prelude to *Tristan*, nascent and growing passion is presented by means of gradually rising pitch, but at the very end of the *Liebestod*, the sense of fulfilment and finality finds expression in falling phrases. . . .'

A number of analyses have appeared, showing the symbolic meanings of the thematic material of J. S. Bach, and how these accord with the words. It is only a pity that some of this writing appears to represent Bach consulting his 'symbolic phrase-book', rather than finding the right phrase for a particular text by allowing his instinctive genius to experiment freely with associations.

The Physical Effects of Music

If music can be identified in the brain with many other processes of the mind and body which have similar patterns, we might expect that for some people, and on some occasions, the effects would go further, and musical patterns would actually stimulate activity in the allied processes. Bodily activities are easier to observe and measure than mental ones, so if there is any evidence of such interaction we should expect to obtain it from the physiologists.

An obvious example of such an effect immediately comes to mind. Just as the ordinary rate of walking influences our ideas of what is 'normal' in the way of tempo, so most people feel the urge to make their feet move in time to an appropriate musical rhythm. Marching and dancing always go better to music, and are less fatiguing with music than without. Foot-tapping or swaying in time to the music are common reactions to strong rhythms.

[1] John Lane, London, 1912.

I am not suggesting that they add to the enjoyment of the music in any great degree, but, in fact, they do no harm: there is a certain degree of snobbishness about this particular reaction to music. Children are taught by well-meaning teachers or parents that it is uncultured or even unmusical to tap or sway to even the most rhythmic and gay 'serious' music, but are expected to admire conductors and performers whose terpsichorean posturing would be more appropriate on the dance-floor than on the platform. It is not necessary to react in this way to music, but it sometimes arises because of an overflow of the musical patterns into the motor side of our organism.

P. E. Vernon,[1] analysing the motor reactions of a mixed audience (university musical instructors, musical amateurs, and so on down to 'unmusical psychologists'—Vernon's rating, not mine) says: 'Rhythm is an aspect of music that is more of a bodily than an auditory nature; over three-quarters of my audience acknowledged making considerable or moderate bodily responses to it, though often only feeling tendencies to move their legs, hands, etc., rather than actually swaying them to and fro. Again the more musical seem to respond less overtly, more mentally or implicitly. We see, therefore, that much that is non-auditory may be quite efficient and helpful in appreciation and, far from being non-musical.' He goes on to say: 'One must allow that these motor responses are capable of evoking the emotion or mood which they characteristically accompany, as in Ribot's classical theory of emotional memory.'

Another familiar type of bodily reaction, apart from tapping or swaying, is the more sophisticated reaction of moving the vocal cords in a silent pantomime of singing, or moving the fingers as if playing an instrument.

Apart from these movements, which are entirely under our control and conscious, it is interesting to see whether music has any effect on the automatic muscular activities such as breathing or heartbeat. A. E. M. Grétry, the Belgian operatic composer, has left a curious little record of what must be one of the first

[1] *Brit. J. Psych.*, vol. 21, 1930, p. 50, and *Musical Times*, 1st March 1929.

serious experiments in this field of investigation: 'I placed three fingers of my right hand on the artery of my left arm, or on any other artery in my whole body, and sang to myself an air the tempo of which was in accordance with my pulse: some little time afterwards, I sang with greater ardour in a different tempo, when I distinctly felt my pulse quickening its action or slackening its action to accommodate itself by degrees to the tempo of the new air.' It is easy to imagine that singing with 'greater ardour' might quicken the pulse through sheer exertion, but the fact that Grétry also mentions a slowing of the pulse is difficult to explain without allowing some sympathetic action between the heartbeat and the music. Later experimenters were able to carry out experiments on other people so as to exclude this obvious source of error in Grétry's 'do-it-yourself' methods.

Binet and Courtier in 1895 found the following change in pulse rate and rate of respiration caused by musical selections of various tempi:

	Pulse/minute		Breathing/minute	
	Before	*After*	*Before*	*After*
March from *Tannhäuser*	80	84	9·6	13·5
Soldiers' March (*Faust*)	81	87	9·0	12·5
Hungarian March (*Damnation of Faust*, Berlioz)	86·5	91·5	11·0	14·5
Sword Episode (*Walküre*)	69	70	9·5	14·5
Ride (*Walküre*)	68	83	9·0	14·0
Spring Song (Wagner)	69	73·5	9·5	13·0
Meeting Scene (*Faust*)	68	84	10·0	13·0
Love Duet (*Faust*)	73	83	10·0	12·0

The effect of the 'Ride' from *Die Walküre* is very marked.

In later experiments with better statistical design, Ellis and his co-workers[1] measured pulse and rate of respiration before and after the performance of 'subdued jazz' (Hall's *Blue Interval*), Debussy's *Prelude à l'Après-Midi d'un Faun*, and Liszt's *Hungarian*

[1] *Amer. J. Psych.*, vol. 65, 1952, p. 39.

Rhapsody No. 3. They found that all three pieces of music resulted in an increased rate of respiration during performance, though the Debussy produced this result very much more slowly than the other two pieces, as might be expected from its broad and slow-moving rhythms. The Liszt, again quite predictably, produced a greater effect than the other two pieces. The rate of respiration, in every case, slowed down soon after the music had finished. Measurement of the heartbeat showed that only the exciting *Hungarian Rhapsody* produced any significant increase in rate.

These were averaged results from a number of students who were not necessarily all musical or even interested in music, and therefore the averages must include some people who were hardly affected at all. I. H. Hyde[1] compared the effects of Tchaikovsky's Sixth Symphony with Sousa's *National Emblem* March. He says: 'On persons not susceptible to music the tragic minor tones that characterized Tchaikovsky's symphony were without effect. But in persons endowed with musical sensitivity the tones of the selections produced a stimulation that as a rule reflexly lowered the functions especially considered in this investigation. . . . As a rule, the records secured on the effect of the *National Emblem* march, showed an increase in the cardiovascular activity, especially noticeable in the velocity of the blood flow and systolic and diastolic blood pressure. But for those who were not able to keep step with the march and lacked fondness for music the records remained unchanged.'

Which goes to show that attention, as we saw in the experiments with cats, is essential for a signal to make any headway in the brain.

For those who are susceptible to music the effects can go a good deal further, if we are to take at their face value the claims made by some writers for music as a therapeutic agent. I. M. Altschuler lists among the physical effects of music the capacity to produce changes in metabolism, respiration, blood pressure, pulse and endocrine and muscular energy. In his work at the

---

[1] 'Effects of Music upon Electrocardiograms and Blood Pressure', in *The Effects of Music*.

Eloise Hospital, Michigan, Altschuler has used music as a soothing agent for mentally distressed patients; to quote his own
words—'music is 35 per cent more effective in quieting disturbed
patients than the wet-pack method'. There can be few more
touching and elegant tributes to St. Cecilia.

However, the use of music in healing has a long and respectable
history, dating back to the legends of Apollo, the original *healer*,
as his name implies, and David playing to the mentally-distressed
Saul. 'Musica mentis medicina moestae' (music is the medicine
of a troubled mind) said Haddon in 1567, and Pope agrees in
the *Ode on St. Cecilia's Day*—'Music can soften pain to ease'.
William James introduced the use of music in mental hospitals,
and it is commonly used now for its therapeutic effects. The
Phipps Psychiatric Clinic of the John Hopkins Hospital, Balti-
more, even has its own music department to improve and refine
the use of music in psychiatry.

These are, of course, mental effects. To return bodily effects,
C. M. Diserens[1] lists the following which can be used for the
therapy of the body:

(1) It increases metabolism.
(2) It increases or decreases muscular energy.
(3) It accelerates breathing and decreases its regularity.
(4) It produces a marked but variable effect on blood volume,
    pulse and blood pressure.
(5) It lowers the threshold for sensory stimuli of different
    modes.

Similar accounts may be found in two more recent books,[2] [3]
dealing with this aspect of music. An excellent summary of
earlier work is given by D. Soibelman.[4]

It should be stressed that the many difficulties in conducting

---

[1] *The Influence of Music on Behaviour*, University Press, Princeton, 1926.
[2] E. Podolsky, *Music for Your Health*, Ackerman, New York, 1945.
[3] P. H. Apel, *The Message of Music*, Vantage Press, New York, 1958.
[4] *Therapeutic and Industrial Uses of Music*, Columbia University Press, New
York, 1948.

experiments of this nature had not been fully realized by all the earlier workers, and their experimental designs leave much to be desired. Too little account was taken of the possible exciting or disturbing effects of the experiments or apparatus. The equipment for taking electrocardiograms, blood pressure, rate of respiration, and so on, is not uncomfortable—but to sit with such attachments, under the eye of a doctor, and to listen to music knowing that some sort of response is expected of you, is a very different thing from sitting in the concert hall or in a comfortable armchair by the gramophone.

The other fault which one must find with much of this work is the trivial nature of the music used in most of the experiments. To compare the patients' reactions to a Sousa march or 'To a Wild Rose' may tell us a little about the effects of fast and slow music respectively, but how much more interesting to compare, for instance, the finale of Beethoven's Seventh Symphony with the allegretto!

However, these objections aside, there seems to be enough information to justify the belief that music has definite bodily effects on susceptible (or just musical?) people, and it is obvious that there is a rewarding study to be carried on, using modern methods and instruments. The elements of musical language, as classified in Chapter Two, and in such works as *The Language of Music*, could also be analysed to see how far their emotional associations correspond to their physical effects. So far, because the suitable instruments are widely distributed, studies have been mainly on pulse, respiration and blood pressure changes, but other bodily reactions, such as the digestive system, the pelvic reflexes, the skin, and so on, would well repay attention, despite the experimental difficulties.

To quote one concrete example: there is a pattern of phrase widely used in music of all periods, but particularly in the Romantic to modern composers, which may be described as an *erotic* one: Deryck Cooke has collected examples of this in *The Language of Music*. Zerlina in *Don Giovanni*, as she presses Masetto's hand to her heart, sings

'Sentilo bat-te-re, Sentilo bat-te-re....'

Tristan, falling in hopeless love with Isolde, is characterized by a similar theme in the Introduction to the First Act

and Tchaikovsky, in the slow movement of the Fifth Symphony, produces an effect of heavy passion by using a similar phrase to that of Zerlina, but in falling sequence instead of a rising one. In each case there is a rising group of notes, the top note repeated and accented on the second repetition, then a fall of a fifth, sixth or seventh. The same phrase, without the repeated climax note or the accent at this point, occurs repeatedly in Tchaikovsky's *Romeo and Juliet* overture and Gershwin's *Rhapsody in Blue* and several other compositions, including popular songs where the phrase is overtly united to erotic words. It would be interesting to see whether this phrase is, as it seems, an expression of *yielding* (rather than assertive) love, as it is certainly meant to be in the music for Zerlina, and probably in the Tchaikovsky example, and also to see whether it has a different significance for men and women.

Applications to Newer Composition

The examples that I have chosen of musical patterns equivalent to mental or physiological patterns have been taken from the established repertoire, roughly from Bach to Ravel. The reasons for this are obvious: the music will easily come to the mind of the reader, and there is little disagreement about the importance of the examples as music.

However, it will be obvious that the same methods of analysis could be applied to music of all periods, and an understanding of the patterns which are most important to the listener could well lead to the use of such patterns in new musical compositions. Even the newest methods of composition, such as the computer progamming of music, need to take account of patterns of tension and resolution, and when the resource of harmonic or melodic tension is absent, as in *musique concrète* and other compositions with 'noises' alone, the patterns derived from rhythm, volume and tone-colour can still be fully exploited.

One valid criticism that can be levelled against the composers of computer-derived music in particular is that they take too little account of these 'human' patterns, and tend to rely on inorganic sources for their material: mathematical series or tables of random numbers, etc. It is quite possible even now to analyse the most important human tension/resolution patterns into numerical terms, and use these as the generating material for a computer programme, and the results are far more likely to be acceptable music.

Similarly the composers of electronic music might consider using the patterns of breathing, heartbeat and other human activities, as their formative elements. By the use of such a synthesizing device as the ring modulator these patterns could be combined with electronic or recorded material.

To summarize this chapter, music can be analysed into patterns of tension and resolution which can be as short as a two-note phrase, or more complex units up to the complete length of a work such as a symphony or opera. These patterns have no meaning for us in themselves, but they can be perceived to be analogous to patterns of tension and resolution which arise from bodily and mental activities. From the point of view of certain regions of the brain, the musical patterns and the 'human' patterns may be so similar that they are identical in basic form, although it is not necessary for every detail to be identical.

In this way music can have a meaning for us, and some personal interest. In the next chapter I have attempted to analyse how these meanings can acquire, in addition, a value.

6 The Levels

In the previous chapters, we have considered the close relationship between musical tension/resolution patterns, and the patterns of mental or bodily activities, and have seen that the particular method by which the brain operates lends itself to a synthesis, blending, or identification of the patterns from different sources, so that there is a mechanism for music to become, in Langer's phrase, 'the analogue of emotive life'. It can also, perhaps more than any of the other arts, be the analogue of bodily life as well.

However, this only takes us one stage in our analysis, and it does not account very satisfactorily for the power of music to move the listener, to enlarge the consciousness. If the function of music were only to simulate the effects of simple emotions, it would not be a very important art, and if it only simulated the effects of heartbeat, breathing and bowel action, it would serve no purpose at all.

The actual effects of great art are far more complex and constitute a synthesis of many emotions and feelings to make up a whole that is greater than the sum of its parts. Edmund Gurney[1] puts the matter very clearly: '... the prime characteristic of Music, the alpha and omega of its essential effect, namely its perpetual production in us of an emotional excitement of a very intense kind, which yet cannot be defined under any known head of emotion. So far as it can be described, it seems like a fusion of strong emotions transfigured into a wholly new experience, whereof if we seek to bring out the separate threads we are hopelessly baulked: for triumph and tenderness, desire

[1] *The Power of Sound*, London, 1880.

and satisfaction, yielding and insistence, may seem to be all there at once, yet without any dubiousness or confusion in the result; or rather elements seem there which we struggle dimly to adumbrate by such words, thus making the experience seem vague only in our own attempt to analyse it, while really the beauty has the unity and individuality pertaining to clear and definite form.'

Gurney's analysis of the effects of music gives us the clue, in fact, to the last stage in our analysis of the methods by which music achieves its effects. The stages might be set out briefly as follows:

(1) Music is made up of patterns of tensions and resolutions;
(2) These patterns correspond to those of activities in the brain caused by mental and bodily events;
(3) The patterns correspond to several different mental and bodily activities, so that the listener is made simultaneously aware of all these activities in a synthesis or fusion. This synthesis constitutes the aesthetic experience.

There is no difficulty about the adequacy of music, as a language, to express several different things simultaneously—to take the simplest type of example, there are countless passages in which the harmonic scheme creates one set of tension/resolution relationships, while the rhythmic scheme creates another set, and a large pattern may include in itself several smaller patterns: the pattern of scale passages mentioned in the last chapter (p. 83) as representative of 'breathing' patterns, and associated with the scene between Giovanni and the statue of the Commendatore, is repeated many times within the larger framework of the harmonic scheme in this passage, which is itself only one of the building units for the whole scene. If we accept that musical patterns are analogues of mental and bodily ones, in fact, it is more likely that several simultaneous patterns should go on than that one simple pattern should be isolated. What is more obscure is the reason why this simultaneity should give us more pleasure than the recreation of a single pattern.

The reason lies in the ability of music, and the other arts, to appeal simultaneously to different *levels* of our personality, so that we are made aware, at one and the same time, of intellectual, emotional and bodily patterns, and we are also shown that many of these patterns have the same basic 'shape'. Not all of the patterns will be associated with conscious experiences: there will also be the analogues of unconscious and forgotten events, and the restatement of these, in a synthesis with conscious ones, can account for the 'mysterious' moving quality of music, and the difficulty that many people find in putting the experience of music into words. Even if we have not tried to find the words ourselves, most of us, at some time or another, have been repelled by other writers' efforts to 'translate' music, for programme notes and so on, and the totally inadequate way in which these attempts compare with the music itself. This failure becomes understandable, and indeed inevitable, if the words have to explain several events at once, including some which are buried in the unconscious.

Some of the unconscious appeal of music will be to the levels of the *personal unconscious* or *individual unconscious* (feelings and emotions which have been forgotten or repressed from the conscious mind) but some of the appeal will be to deeper levels, the levels of common human experience and innate emotions which some writers would call instinct, and which Jung called the *collective unconscious*. Jung defines it as follows: 'While the personal unconscious is made up essentially of contents which have at one time been conscious but which have disappeared from consciousness through having been forgotten or repressed, the contents of the collective unconscious have never been in consciousness, but owe their existence exclusively to heredity. Whereas the personal unconscious consists for the most part of *complexes*, the content of the collective unconscious is made up essentially of *archetypes*.'[1] One does not have to accept all of Jung's detailed theory about the functioning of the collective unconscious to see in his work a detailed analysis of the effects on human behaviour

[1] *The Concept of the Collective Unconscious*, Collected Works, vol. 9, tr. Hebb, Routledge and Kegan Paul, London, 1959.

of those common experiences, birth, feeding, excreting, sleeping, and their patterns, and also the instinctual patterns of behaviour. His main interest was the way in which the contents of the collective unconscious, often too unpleasant to be allowed naked into our unconsciousness, are clothed in the trappings of myth and folklore: in this form they are the province of poetry and other literature, and much of plastic art. The approach of music to this repository of instinct is to recall the shapes and patterns, almost the diagrams, of the archetypes, and to bring them into consciousness in this form. In either case, a synthesis takes place.

Jung himself discusses the effects of appealing simultaneously to both levels of the mind, conscious and unconscious, and points out that both must be allowed to function and interact to make a full personality. As he says:[1] 'Conscious and unconscious do not make a whole when one of them is suppressed and injured by the other. If they must contend, let it at least be a fair fight with equal rights on both sides. Both are aspects of life. Consciousness should defend its reason and protect itself, and the chaotic life of the unconscious should be given the chance of having its way too—as much of it as we can stand. This means open conflict and open collaboration at once. That, evidently, is the way human life should be. It is the old game of hammer and anvil: between them the patient iron is forged into an indestructible whole, an "individual".'

When collaboration occurs, when, for a while, the lines of conscious and unconscious thought run along the same track, we achieve the feeling of wholeness and satisfaction which is characteristic of our response to great art and other transcendent states of mind. The patterns of music, translated, analysed, shorn of detail, are able to simulate the patterns of emotions on many levels simultaneously, thus bringing various hierarchical states of consciousness and unconsciousness into harmony with one another during the existence of the music for us, whether this is in a performance or purely in the memory.

[1] *Conscious, Unconscious and Individuation*, Collected Works, vol. 9, tr. Hebb, Routledge and Kegan Paul, London, 1959.

As this happens we experience the sense of unity which arises from the cessation of conflict between conscious and unconscious. And because we are able to detect in music analogies to the deepest and least individual layers of the mind, we become aware also of our unity with the rest of mankind: 'A man to be greatly good must imagine intensely and comprehensively; he must put himself in the place of another and of many others; the pains and pleasures of his species must become his own' says Shelley in *The Defence of Poetry*, and it is this reminder that we belong to a species that is one of the enlightenments of great art. Breathing, feeding, loving, grief, frustration and happiness are the common lot of mankind. Man is the only species for whom symbols are significant, and the symbols which can be created by music and the other arts can comprehend these basic human experiences simultaneously with other, more detailed, more conscious experiences. To compare the basic patterns as they appear in the brain, in fact, the symbol and the reality, the conscious reaction and the unconscious one, are the same, since the same pattern holds for all of them.

The greatest art meets us on the greatest number of levels, and conversely, poverty in art, whether we are considering the theatre, music, or any other means of communication, arises from the failure to communicate on more than one or two levels. Poets use the techniques of ambiguity and multiple meanings, as well as the musical techniques of rhythm and sound, to create this approach on many levels, so that *Macbeth*, for example, is at once an eleventh-century blood-and-thunder thriller, an allegory of the corrupting influence of ambition and power, and a distillation of all the night-fears of humanity in one vast storm-shaken nightmare. It is also other things for which there are no clear words. The important thing about the multi-level approach is that a reconciliation or harmonization should take place between the levels of the mind that are, most of the time, operating independently or even in conflict. Kant emphasizes this 're-conciliation' aspect of beauty in the *Critique of Judgement*, pointing out that beauty tends to stimulate our faculties of imagination

(*Vorstellung*) and understanding (*Erkenntnis*) in 'harmonious interaction', and ascribes pleasure in art to this reconciliation of these often divergent parts of the personality.

More recently, Koestler[1] has developed at length a theory of artistic and scientific creativity which provides an interesting parallel to the theory of artistic appreciation outlined in this chapter. His basic point is that the conditions for true creation, whether in the form of a work or art or of a fruitful scientific concept, are set up when an idea or external percept is suddenly seen to fit into two systems or planes of thought which seemed previously to have little or nothing in common. Koestler discusses many examples, particularly in *The Act of Creation*, including the classical scientific discoveries of Poisson, the discovery of movable type by Gutenberg, Darwin's realization of the process of natural selection and Archimedes's recognition that the shift in the level of his bath water gave the essential clue to the measurement of the volume, and hence the density, of any complex shape, whether that of the human body or that of Hieron's crown. Because of this early and well-known example, Koestler sometimes refers to the discovery as the *Eureka-act* but his coined word *bisociation* is a convenient shorthand for the process of fusing or superimposing different levels of consciousness or frames of mental reference.

The idea of artistic communication on many levels simultaneously is, in fact, one that has been adumbrated by many writers. William Empson[2] has explored in detail the use of ambiguities in poetry so as to express more than one meaning, usually a surface meaning and one or more deeper ones. It is hardly necessary to quote from his analyses, but in the preface to the second edition of his book he has expressed his basic theory with habitual lucidity: '. . . whenever a receiver of poetry is seriously moved by an apparently simple line, what are moving in him are the traces of a great part of his past experience and of

---

[1] *Insight and Outlook*, Macmillan, London, 1949, and *The Act of Creation*, Hutchinson, London, 1964.

[2] *Seven Types of Ambiguity*, second edition, Peregrine Books, London, 1961.

the structure of his past judgements.' And, a little earlier in the same section: '. . . there is always in great poetry a feeling of generalisation from a case which has been presented definitely; there is always an appeal to a background of human experience which is all the more present when it cannot be named.'

Not only in poetry and music, but in all the arts, it is intuitively acceptable (and accepted by many writers of all periods) that the multiple-level character of great art accounts for its numinous quality, and, as we have seen, the mechanisms of the brain make communication on many levels not only possible, but probable. Given the tendency of the brain to reduce events to their basic patterns, for the process of identification (a process conveniently, if inelegantly, called *departicularization* by most writers on the subject) it is intrinsically more likely that a complex pattern such as a piece of music or poetry will recall many other patterns, than that it will correspond to only one. Every time we receive a sensory impression, or recall an earlier one, we have to go through this process of stripping off the superficial details of the pattern and comparing it with other patterns. We are searching for analogies and similarities. The fact that the process is carried out by translation of all the patterns into an electrical pulse code is merely a detail: if the brain were constructed to work on some other system (the variable voltage of the analogue computer, for example, or the pressure of a fluidics system) it would still have to develop some method of departicularization for sifting out the 'shapes' of experience.

Of course, we can take a simple pleasure in single levels of art: pretty colours or patterns, pleasant noises, colourful words, and so on, but such material does not have the same sense of combined mystery and satisfaction as multiple-level art. It would be convenient if there were some accepted word for such superficial qualities: if 'beauty' is represented by multiple-level productions, then 'concinnity' is a reasonable name for the single-level material. Many *objets trouvés* are concinnous, like a finely-marked stone or the sounds of an aeolian harp, and many of the productions of 'computer art' are at the moment in this class.

No investigations of the brain have yet gone far enough to find out in detail what qualities in the multi-level patterns account for the great pleasure and intellectual satisfaction that they can bring us, but it will be possible to analyse this in the future. This is a scientific problem of the type that does not need a flash of inspiration for its solution, only the time and resources. For now, at least, we can see that certain patterns have the effect of calling up recollections of many similar patterns from past experiences and that these recollections arise from many different storage areas of the brain and from many different levels of the personality from the most rarefied heights of self-awareness to the darkest depths of the unconscious. Similarities in these patterns are then drawn to our attention by their juxtaposition, and we find ourselves experiencing a synthesis or fusing of many events, many memories, many of the paradigms of existence. This is in itself a new experience, and one which is very much more profound and stirring than the individual experiences of which it is composed. It is, in truth, a mystical experience, and we might say of it what William James said of the religious mystical state: 'The keynote of it is invariably a reconciliation. It is as if the opposites of the world, whose contradictoriness and conflict make all our difficulties and troubles, were melted into unity.'[1]

[1] *The Varieties of Religious Experience*, Longmans, London, 1902.

Index